Thomas Raskop

Oblique Boundary Value Problems and Limit Formulae of Potential Theory

Thomas Raskop

Oblique Boundary Value Problems and Limit Formulae of Potential Theory

An Analysis Motivated by Geomathematical Problems

Südwestdeutscher Verlag für Hochschulschriften

Impressum/Imprint (nur für Deutschland/ only for Germany)
Bibliografische Information der Deutschen Nationalbibliothek: Die Deutsche Nationalbibliothek verzeichnet diese Publikation in der Deutschen Nationalbibliografie; detaillierte bibliografische Daten sind im Internet über http://dnb.d-nb.de abrufbar.

Alle in diesem Buch genannten Marken und Produktnamen unterliegen warenzeichen-, marken- oder patentrechtlichem Schutz bzw. sind Warenzeichen oder eingetragene Warenzeichen der jeweiligen Inhaber. Die Wiedergabe von Marken, Produktnamen, Gebrauchsnamen, Handelsnamen, Warenbezeichnungen u.s.w. in diesem Werk berechtigt auch ohne besondere Kennzeichnung nicht zu der Annahme, dass solche Namen im Sinne der Warenzeichen- und Markenschutzgesetzgebung als frei zu betrachten wären und daher von jedermann benutzt werden dürften.

Verlag: Südwestdeutscher Verlag für Hochschulschriften Aktiengesellschaft & Co. KG
Dudweiler Landstr. 99, 66123 Saarbrücken, Deutschland
Telefon +49 681 37 20 271-1, Telefax +49 681 37 20 271-0
Email: info@svh-verlag.de
Zugl.: Kaiserslautern, TU, Diss., 2009

Herstellung in Deutschland:
Schaltungsdienst Lange o.H.G., Berlin
Books on Demand GmbH, Norderstedt
Reha GmbH, Saarbrücken
Amazon Distribution GmbH, Leipzig
ISBN: 978-3-8381-1329-6

Imprint (only for USA, GB)
Bibliographic information published by the Deutsche Nationalbibliothek: The Deutsche Nationalbibliothek lists this publication in the Deutsche Nationalbibliografie; detailed bibliographic data are available in the Internet at http://dnb.d-nb.de.

Any brand names and product names mentioned in this book are subject to trademark, brand or patent protection and are trademarks or registered trademarks of their respective holders. The use of brand names, product names, common names, trade names, product descriptions etc. even without a particular marking in this works is in no way to be construed to mean that such names may be regarded as unrestricted in respect of trademark and brand protection legislation and could thus be used by anyone.

Publisher: Südwestdeutscher Verlag für Hochschulschriften Aktiengesellschaft & Co. KG
Dudweiler Landstr. 99, 66123 Saarbrücken, Germany
Phone +49 681 37 20 271-1, Fax +49 681 37 20 271-0
Email: info@svh-verlag.de

Printed in the U.S.A.
Printed in the U.K. by (see last page)
ISBN: 978-3-8381-1329-6

Copyright © 2010 by the author and Südwestdeutscher Verlag für Hochschulschriften Aktiengesellschaft & Co. KG and licensors
All rights reserved. Saarbrücken 2010

Für

Johann Friedrich Raskop

und

Peter Jagiella

DANKSAGUNG

Als erstes möchte ich mich bei Prof. Dr. Martin Grothaus für die Betreuung meiner Promotion bedanken. Ohne sein Engagement wäre diese Arbeit weder fachlich noch finanziell realisierbar gewesen. Besonders bedanken möchte ich mich auch für die eingeräumten Freiräume und das damit verbundene Vertrauen. Im gleichen Atemzug möchte ich mich bei Prof. Dr. Willi Freeden bedanken. Wann immer ich eine Frage hatte, stand seine Tür offen und durch viele interessante, inspirierende Diskussionen und Anregungen hat er erheblichen Anteil an den erreichten Ergebnissen. Desweiteren möchte ich mich bei Prof. Dr. Heinrich von Weizsäcker für wertvolle Anmerkungen und Kommentare während meiner Vorträge, sowie bei Dr. Martin Gutting für seine freundliche Hilfe bei fachlichen Fragen bedanken. Ein weiteres Dank geht an die Mitglieder der Arbeitsgruppe Funktionalanalysis, Anna Vogel, Lama Osman, Dr. Torben Fattler, Florian Conrad, Wolfgang Bock und Thanh Tan Mai, sowie Frau Cornelia Türk und Frau Claudia Korb. Für die angenehme Arbeitsatmosphäre und die Hilfsbereitschaft bei Problemen aller Art. Auch bei Manar Omari möchte ich mich für die Diskussionen, im Zusammenhang mit der Betreuung ihrer Masterarbeit, bedanken. Meine Dankesrunde im Bereich der Universität möchte ich nicht zuletzt mit einem Dank an den Fachbereich Mathematik sowie an die Landesgraduiertenförderung der TU Kaiserslautern beenden. Ohne ihre finazielle Unterstützung wäre meine Promotion unmöglich gewesen.

Weiter geht es mit meinen ehemaligen WG-Mitbewohnern Eva Jennewein, Matthias Klug und Mr. Wankel. Vielen Dank für die schöne Zeit. Insgesamt möchte ich mich bei all meinen Freunden bedanken. Vielen Dank für die sieben Jahre mit euch in Kaiserslautern, in denen wir viel erlebt haben und einigen Spaß hatten. Namentlich nennen möchte ich Paul Fischer, Jochen Hoffmann, Stefan Hopf und Florian Lehnhardt.

Das beste kommt zum Schluss, so ist es auch in dieser Danksagung. Ein großes Dankeschön geht an meine Familie, Annegret, Anna, Maria und Stefan Raskop für die Unterstützung in jeder denkbaren Art. Der letzte Dank gilt meiner Freundin Dominika Jagiella. Danke das du immer für mich da warst. Zurückblickend warst du das, wofür es sich am meisten gelohnt hat nach Kaiserslautern zu kommen.

Contents

1 Introduction   **9**
   1.1   The Oblique Boundary Problem for
        the Poisson Equation   9
   1.2   Limit Formulae and Jump Relations of
        Potential Theory   14

2 Preliminaries   **21**
   2.1   Domains and Surfaces   21
   2.2   Function Spaces   26
   2.3   Properties of Sobolev Spaces on Submanifolds   37
   2.4   Tools from (Functional) Analysis   40
   2.5   Some Notions from Probability Theory   44
   2.6   Function Systems from Geomathematics   45

3 Oblique Boundary Problems   **47**
   3.1   The Inner Regular Oblique Boundary Problem   47
   3.2   Transformations   53
       3.2.1   Kelvin Transformation of the Domain   54
       3.2.2   Kelvin Transformation of the Solution   57
       3.2.3   Transformation of Inhomogeneities and Coefficients   60
   3.3   The Outer Oblique Boundary Problem of Potential Theory   69
       3.3.1   Weak Solutions to the Outer Problem   69
       3.3.2   The Condition on the Oblique Vector Field   74
       3.3.3   Stochastic Inhomogeneities   76
       3.3.4   Ritz-Galerkin Approximation   78
       3.3.5   Geomathematical Applications and Examples   80

4 Limit Formulae and Jump Relations — 83
- 4.1 Definition and Properties of the Layer Potentials — 83
- 4.2 Pointwise and Uniform Convergence — 85
- 4.3 Limit Formulae in $C^m(\partial\Sigma)$ — 88
- 4.4 Limit Formulae in Hölder Norms — 92
- 4.5 Limit Formulae in $L^2(\partial\Sigma)$ — 98
- 4.6 Limit Formulae in Sobolev Spaces — 100
- 4.7 Application of the Limit Formulae to Geomathematics — 112
- 4.8 Appendix: Oblique Limit Formulae — 122

Chapter 1

Introduction

This dissertation deals with two main subjects. Both are strongly related to boundary problems for the Poisson equation and the Laplace equation, i.e., the homogeneous Poisson equation. The oblique boundary problem of potential theory is treated in Chapter 3, while the limit formulae and jump relations of potential theory are investigated in Chapter 4. Consequently we divide the introduction into two parts.

1.1 The Oblique Boundary Problem for the Poisson Equation

The main subject of Chapter 3 are existence results for solutions to the outer oblique boundary problem for the Poisson equation. It is based on the article [GR06], in which a theory for deterministic as well as stochastic inhomogeneities and solutions to the regular inner problem is provided. The problem is called outer problem because it is defined on an outer domain $\Sigma \subset \mathbb{R}^n$. This is a domain Σ, having the representation $\Sigma = \mathbb{R}^n \backslash \overline{D}$ where $0 \in D$ is a bounded domain. Consequently, $\partial \Sigma$ divides the euclidean space $\mathbb{R}^n$ into a bounded domain D, called inner domain, and an unbounded domain Σ, called outer domain. A problem defined on D and ∂D is called inner problem. The Poisson equation in the domain is given by

$$\Delta u = f,$$

and the oblique boundary condition by

$$\langle \underline{a}, \nabla u \rangle + bu = g.$$

This condition is called regular if the equation

$$|\langle \underline{a}, \nu \rangle| > C > 0,$$

holds on $\partial \Sigma$, or ∂D respectively, for a constant $0 < C < \infty$. A classical solution corresponding to continuous $\underline{a}$, b, g and f of the outer oblique boundary problem for the Poisson equation is a function $u \in C^2(\Sigma) \cap C^1(\overline{\Sigma})$ which fulfills the first two equations and is regular at infinity, i.e., $u(x) \to 0$ for $|x| \to \infty$. Existence and uniqueness result for a classical solution to the regular outer oblique boundary problem for the Poisson equation are already available, see e.g. [Mir70, Section 23]. In order to allow very weak assumptions on boundary, coefficients and inhomogeneities, we are interested in weak solutions. Due to our technique, we do not have to consider a regular outer problem. In [GR06] an existence and uniqueness result for the weak solution to the inner regular oblique boundary problem is presented. To be more precisely, if $\partial \Sigma$ is a $C^{1,1}$-boundary of a bounded domain Σ, $|\langle \underline{a}, \nu \rangle| > C > 0$ on $\partial \Sigma$, $\frac{\underline{a}}{\langle \underline{a},\nu \rangle} - \nu \in H^{1,\infty}(\partial \Sigma; \mathbb{R}^n)$, $\frac{b}{\langle \underline{a},\nu \rangle} \in L^\infty(\partial \Sigma)$ and ess $\inf_{\partial \Sigma} \left(\frac{b}{\langle \underline{a},\nu \rangle} - \frac{1}{2}\text{div}_{\partial \Sigma}(\frac{\underline{a}}{\langle \underline{a},\nu \rangle} - \nu) \right) > 0$, then for every $\frac{g}{\langle \underline{a},\nu \rangle} \in H^{-\frac{1}{2},2}(\partial \Sigma)$ and $f \in (H^{1,2}(\Sigma))'$ there exists one and only one $u \in H^{1,2}(\Sigma)$ fulfilling the weak formulation. Additionally the solution depends continuously on the data. Beside from some results for Sobolev spaces defined on submanifolds, the crucial point in the proof is to show coercivity of the bilinear form related to the weak formulation. Therefore we use a Poincaré inequality, namely

$$\int_\Sigma \langle \nabla u, \nabla u \rangle \, d\lambda^n + \int_{\partial \Sigma} u^2 dH^{n-1} \geq C \left(\int_\Sigma u^2 \, d\lambda^n + \int_\Sigma \langle \nabla u, \nabla u \rangle \, d\lambda^n \right),$$

for all $u \in H^{1,2}(\Sigma)$. For details see the reference given above. Important is, that this Poincaré inequality is only available for bounded domains, so we can not apply the same techniques to the outer setting. Before we go to the outer problem we prove a regularization result for the inner problem. This well be used later on in order to prove the main results of this article. If $\partial \Sigma$ is a $C^{2,1}$-boundary of a bounded domain Σ, $\frac{\underline{a}}{\langle \underline{a},\nu \rangle} - \nu \in H^{2,\infty}(\partial \Sigma; \mathbb{R}^n)$, $\frac{b}{\langle \underline{a},\nu \rangle} \in H^{1,\infty}(\partial \Sigma)$, $\frac{g}{\langle \underline{a},\nu \rangle} \in H^{\frac{1}{2},2}(\partial \Sigma)$ and $f \in L^2(\Sigma)$, we are able to show that the weak solution is even a strong solution, i.e., $u \in H^{2,2}(\Sigma)$. In the proof we show that the weak solution fulfills the requirements of a regularization result for the weak Neumann problem, taken from [Dob06].

Then we tackle the outer problem. Our approach in order to provide a weak solution is transforming this problem to a corresponding inner problem, using the Kelvin transformation. This transformation defines for each outer domain Σ an inner domain Σ^K via

$$\Sigma^K := \left\{ \frac{x}{|x|^2} \Big| x \in \Sigma \right\} \cup \{0\}.$$

1.1. THE OBLIQUE BOUNDARY PROBLEM FOR THE POISSON EQUATION

In turn, we get for each function u defined on Σ^K a function v by

$$v(x) := \frac{1}{|x|^{n-2}} u\left(\frac{x}{|x|^2}\right),$$

for all $x \in \Sigma$. The first transformation leaves the regularity of $\partial \Sigma$ invariant, while the second has the important property

$$\Delta v(x) = \frac{1}{|x|^{n+2}} (\Delta u)\left(\frac{x}{|x|^2}\right), \quad x \in \Sigma,$$

for all $u \in C^2(\Sigma^K)$. Our idea is to use this transformations in order to provide a weak solution. We transform the outer problem into a corresponding inner problem, then solve this problem and finally transform the weak solution of the inner problem to a function defined in the outer domain. The transformations $T_1(f)$ and $T_2(g)$ of the inhomogeneities as well as $T_3(\underline{a})$ and $T_4(b)$ of the coefficients can be identified by standard calculus. The problem is to find the right function spaces and then to extend the transformations to these spaces. In order to identify them we have to take care of two main aspects. First, the image spaces under the transformations have to fulfill the requirements of the existence and uniqueness result for the regular inner problem. Otherwise we cannot apply the solution operator for the inner problem. Second, the transformations should be continuous. Otherwise the weak solution of the outer problem will not depend continuously on the inhomogeneities. We are able to show that the spaces $H^{-\frac{1}{2},2}(\partial \Sigma)$ for the boundary inhomogeneity, $\left(H^{1,2}_{|x|^2,|x|^3}(\Sigma)\right)'$ for the domain inhomogeneity and $H^{1,2}_{\frac{1}{|x|^2},\frac{1}{|x|}}(\Sigma)$ for the weak solution, are a suitable choice. Here we have Sobolev spaces equipped with weighted Lebesgue measures. Under this conditions we are able to prove that if Σ is an outer $C^{1,1}$-domain, $\underline{a} \in H^{1,\infty}(\partial \Sigma; \mathbb{R}^n)$ and $b \in L^\infty(\partial \Sigma)$, fulfilling $|\langle (T_3(\underline{a}))(y), \nu^K(y)\rangle| > C > 0$ and ess $\inf_{\partial \Sigma^K} \left\{ \frac{T_4(b)}{\langle T_3(\underline{a}), \nu^K \rangle} - \frac{1}{2}\text{div}_{\partial \Sigma^K}\left(\frac{T_3(\underline{a})}{\langle T_3(\underline{a}), \nu^K\rangle} - \nu^K\right)\right\} > 0$, for each $g \in H^{-\frac{1}{2},2}(\partial \Sigma)$ and $f \in \left(H^{1,2}_{|x|^2,|x|^3}(\Sigma)\right)'$, there exists $u := K(S^{in}_{T_3(\underline{a}),T_4(b)}(T_1(f), T_2(g)))$ with $u \in H^{1,2}_{\frac{1}{|x|^2},\frac{1}{|x|}}(\Sigma)$ fulfilling the continuity estimate

$$\|u\|_{H^{1,2}_{\frac{1}{|x|^2},\frac{1}{|x|}}(\Sigma)} \leq C \left(\|f\|_{\left(H^{1,2}_{|x|^2,|x|^3}(\Sigma)\right)'} + \|g\|_{H^{-\frac{1}{2},2}(\partial \Sigma)} \right).$$

This continuity estimate enables us to provide a Ritz-Galerkin method later on. Furthermore we can show that if Σ is an outer $C^{2,1}$-domain and $\underline{a} \in H^{2,\infty}(\partial \Sigma; \mathbb{R}^n)$ and $b \in H^{1,\infty}(\partial \Sigma)$, $f \in L^2_{|x|^2}(\Sigma)$ and $g \in H^{\frac{1}{2},2}(\partial \Sigma)$ we have $u \in H^{2,2}_{\frac{1}{|x|^2},\frac{1}{|x|},1}(\Sigma)$ for the weak solution to the outer problem. Additionally, it fulfills the classical formulation almost everywhere and a corresponding continuity estimate holds. Because of the Kelvin transformation we get a transformed non admissible direction for the oblique vector field $\underline{a}$. For $\mathbb{R}^2$ we can explicitly calculate this direction. It only depends on the geometry of the surface $\partial\Sigma$.

In Chapter 2 we define the function spaces we will use in Chapter 3. This are mainly the spaces of smooth functions and spaces of weakly differentiable functions. We introduce Sobolev spaces on submanifolds and Sobolev spaces equipped with weighted Lebesgue measures. Furthermore, we present some important results about these spaces. Chapter 3 is organized as follows. In Section 3.1 we present the weak theory for the regular inner problem including an existence and uniqueness result for a broad class of inhomogeneities. Except of the regularization result at the end of the section, this are mainly the results contained in my diploma thesis, published in [Ada75]. In Section 3.2 we start with the investigation of the outer problem. We introduce transformations which will be used in order to transform the outer problem to a corresponding inner problem. Also some important properties of those transformations are proved. They will be important in Section 3.3. Here we state the outer problem and in the following we will be able to prove the existence of a weak solution for a very general class of inhomogeneities. The modified regularity condition on the oblique vector field, which occurs because of the Kelvin transformation is investigated separately in Subsection 3.3.2. Finally, we state some results about stochastic inhomogeneities and a Ritz-Galerkin approximation method in Subsection 3.3.3 and Subsection 3.3.4, respectively. Both are implemented, using the techniques and results from [GR06]. The applicability to problems from geomathematics is shown in Subsection 3.3.5.

Our analysis of the outer problem is motivated by problems from geomathematics. Here oblique boundary problems arise frequently, because in general the normal of the Earth's surface does not coincide with the direction of the gravity vector. Therefore, the oblique boundary condition is more suitable then a Neumann boundary condition. For details see [Bau04] or [Gut08].

The main progress achieved in Chapter 3 can be summarized by the following core results:

1. A regularization result for a strong solution to the regular inner problem, i.e., $u \in H^{2,2}(\Sigma)$, is proved, see Theorem 3.1.6.

2. The transformation of the outer oblique boundary problem for the Poisson equation to a corresponding inner problem is provided. Important properties of this transformation are proved, see Lemmata 3.2.3, 3.2.7, 3.2.11, 3.2.13 and 3.2.14.

1.1. THE OBLIQUE BOUNDARY PROBLEM FOR THE POISSON EQUATION 13

3. The existence of a weak solution to the outer problem under weak assumptions on coefficients and surface for a large class of inhomogeneities is proved, see Theorem 3.3.2.

4. An existence result for a strong solution under additional regularity assumptions is proved and the connection to the classical problem is established, see Theorem 3.3.4.

5. The transformed condition on the oblique vector field is investigated, see Subsection 3.3.2.

6. Stochastic inhomogeneities as well as an existence result for the stochastic weak solution are implemented. Additionally a Ritz-Galerkin approximation method for numerical computations is provided. Moreover, the results are applicable to problems from Geomathematics. These results are contained in Subsections 3.3.3-3.3.5.

Outlook

In Chapter 3 we prove the existence of a weak solution to the outer oblique boundary problem for the Poisson equation. Therefore we introduce several transformations. We prove for the transformation of the space inhomogeneity f

$$T_1 : \left(H^{1,2}_{|x|^2,|x|^3}(\Sigma)\right)' \to \left(H^{1,2}(\Sigma^K)\right)'.$$

This transformation is not bijective, i.e.,

$$T_1\left(\left(H^{1,2}_{|x|^2,|x|^3}(\Sigma)\right)'\right) \neq \left(H^{1,2}(\Sigma^K)\right)'.$$

Finding a Hilbert space V, such that the transformation

$$T_1 : V \to \left(H^{1,2}(\Sigma^K)\right)',$$

is bijective would lead to the existence of a weak solution for a even larger class of inhomogeneities. Moreover we have for the transformation K of the weak solution to the inner problem

$$K : H^{1,2}(\Sigma^K) \to H^{1,2}_{\frac{1}{|x|^2},\frac{1}{|x|}}(\Sigma),$$

where we have again

$$K\left(H^{1,2}(\Sigma^K)\right) \neq H^{1,2}_{\frac{1}{|x|^2},\frac{1}{|x|}}(\Sigma).$$

Finding a Hilbert space W such that

$$K : H^{1,2}(\Sigma^K) \to W,$$

is bijective, would give us uniqueness of the solution and more detailed information about the behavior of u and its weak derivatives, when x is tending to infinity. Additionally, we would be able to define a bijective solution operator for the outer problem. This could be used to find the right Hilbert spaces, such that a Poincaré inequality is available. Consequently the Lax-Milgram Lemma would be applicable directly to a weak formulation for the outer setting, which can be derived similar to the inner problem. Then we might have to consider a regular outer problem, because the tangential direction is forbidden for the oblique vector field, if we want to derive a weak formulation, see Section 3.1. In turn we get rid of the transformed regularity condition on $\underline{a}$, investigated in Subsection 3.3.2. The results achieved by this dissertation are then still an alternative in order to get weak solutions for tangential $\underline{a}$. Moreover, the availability of a Poincaré inequality would lead to existence results for weak solutions to a broader class of second order elliptic partial differential operators in outer domains. See [Alt02, Chapter 4] for such second order elliptic partial differential operators for inner domains. Another part that could be a subject of further investigations are the conditions on the coefficients of the boundary condition, i.e.,

$$\operatorname{ess\,inf}_{\partial \Sigma^K} \left\{ \frac{T_4(b)}{\langle T_3(\underline{a}), \nu^K \rangle} - \frac{1}{2} \operatorname{div}_{\partial \Sigma^K} \left(\frac{T_3(\underline{a})}{\langle T_3(\underline{a}), \nu^K \rangle} - \nu^K \right) \right\} > 0,$$

as well as the modified regularity condition on $\underline{a}$, i.e.,

$$\left| \langle (T_3(\underline{a}))(y), \nu^K(y) \rangle \right| > C > 0.$$

Moreover, a generalization to other boundary conditions, e.g. Dirichlet, Neumann or Robin boundary conditions, might be implemented if desired. Finally, a generalization to surfaces with more complex geometry, e.g. non connected domains, non connected boundaries or outer domains Σ containing the origin, might be possible.

1.2 Limit Formulae and Jump Relations of Potential Theory

In Chapter 4 we prove the convergence of the limit formulae and jump relations of potential theory in several norms. More precisely, we investigate the potential of the single layer U_1 and

1.2. LIMIT FORMULAE AND JUMP RELATIONS OF POTENTIAL THEORY

the potential of the double layer U_2, defined by

$$U_1[F](x) := \int_{\partial \Sigma} F(y) \frac{1}{|x-y|} dH^2(y),$$

$$U_2[F](x) := \int_{\partial \Sigma} F(y) \frac{\partial}{\partial \nu(y)} \frac{1}{|x-y|} dH^2(y),$$

for all $x \in \mathbb{R}^3 \setminus \partial \Sigma$, where Σ is an outer domain in $\mathbb{R}^3$ and F is a given function on $\partial \Sigma$, called layer function. An outer domain Σ divides $\mathbb{R}^3$ into a bounded connected inner domain D and an unbounded connected outer domain $\mathbb{R}^3 \setminus \overline{D} = \Sigma$. Furthermore, we investigate the first order normal derivatives of U_1 and U_2. The limit formulae describe the behavior of these potentials when approaching the surface $\partial \Sigma$. In general the limit formulae are different when approaching either from the inner or the outer space. These circumstance is called jump relation and can be obtained by the results for the limit formulae. Therefore we restrict ourselves to the limit formulae, while the jump relations can be computed easily by taking either the difference or the sum of the corresponding limit formula. Note that U_1 and U_2 are analytic, harmonic functions as well in the inner as in the outer space, even for outer C^2-domains Σ and continuous F. Non the less we need additional regularity assumptions in order to prove that the limits, when approaching to the surface, exist. We need the potentials to be Hölder continuous, then we know that a unique Hölder continuous continuation onto $\partial \Sigma$ exists. If we want to prove the convergence in $C^m(\partial \Sigma)$-norm, also the derivatives have to be Hölder continuous. These properties can be ensured for sufficiently smooth surface $\partial \Sigma$ and F. From literature we know that for outer C^2-domains Σ and continuous F we have

$$\lim_{\tau \to 0^+} U_1[F](x \pm \tau \nu(x)) = U_1[F](x), \quad \forall x \in \partial \Sigma,$$

$$\lim_{\tau \to 0^+} \frac{\partial U_1}{\partial \nu}[F](x \pm \tau \nu(x)) = \frac{\partial U_1}{\partial \nu}[F](x) \mp 2\pi F(x), \quad \forall x \in \partial \Sigma,$$

$$\lim_{\tau \to 0^+} U_2[F](x \pm \tau \nu(x)) = U_2[F](x) \pm 2\pi F(x), \quad \forall x \in \partial \Sigma,$$

$$\lim_{\tau \to 0^+} \frac{\partial U_2}{\partial \nu}[F](x \pm \tau \nu(x)) = \frac{\partial U_2}{\partial \nu}[F](x), \quad \forall x \in \partial \Sigma.$$

uniformly in $x \in \partial \Sigma$. The first three formulae can be found in [FM04], while we had to prove the last formula, using results from [CK83] and [Sch31b]. In this case we have to assume $F \in C^{1,\alpha}(\partial \Sigma)$. We can even prove that this convergence holds in $C^{0,\beta}(\partial \Sigma)$-norm, provided the additional assumptions

$$F \in C^{0,\alpha}(\partial \Sigma) \quad \text{for formula 1,}$$

$$F \in C^{0,\alpha}(\partial \Sigma) \quad \text{for formulae 2 and 3,}$$

are fulfilled, where $0 \leq \beta < \alpha \leq 1$. Moreover, we are able to prove that these formulae stay valid in $C^m(\partial\Sigma)$-norm for an outer $C^{m+1,\alpha}$-domain Σ and in $C^{m,\beta}(\partial\Sigma)$-norm for an outer C^{m+2}-domain Σ if

$$F \in C^{m-1,\alpha}(\partial\Sigma) \quad \text{for formula 1,}$$
$$F \in C^{m,\alpha}(\partial\Sigma) \quad \text{for formulae 2 and 3,}$$
$$F \in C^{m+1,\alpha}(\partial\Sigma) \quad \text{for formula 4,}$$

where $m \geq 1$ and $0 \leq \beta < \alpha \leq 1$. In the proofs we mainly use results, taken from [Gün57]. The convergence in $C^m(\partial\Sigma)$-norm is basic for the convergence in the Sobolev spaces $H^{m,2}(\partial\Sigma)$. Another result, which is essential in the proof of our main result is the convergence of U_1, U_2 in $L^2(\partial\Sigma)$-norm proved in [Ker80], [Geh70] and [Fre80]. We have this convergence for each outer C^2-domain Σ and each $F \in L^2(\partial\Sigma)$. The main result of the article is the following. We prove the convergence of the formulae above in $H^{m,2}(\partial\Sigma)$ under the conditions

$$F \in H^{m,2}(\partial\Sigma) \text{ and } \Sigma \text{ an outer } C^{m+1,\alpha}\text{-domain for formula 1,}$$
$$F \in H^{m+1,2}(\partial\Sigma) \text{ and } \Sigma \text{ an outer } C^{m+2}\text{-domain for formulae 2 and 3,}$$
$$F \in H^{m+2,2}(\partial\Sigma) \text{ and } \Sigma \text{ an outer } C^{m+3}\text{-domain for formula 4.}$$

Therefore we use the BLT Theorem, which provides us an unique extension of the potential operators from $C^m(\partial\Sigma)$ onto $H^{m,2}(\partial\Sigma)$, provided we are able to prove that these operators are continuous in $H^{m,2}(\partial\Sigma)$-norm on that dense subset. Therefore we estimate the potential operators on parallel surfaces $\partial\Sigma^{\pm\tau}$ by a constant independent of $\tau > 0$. The advantage is that we avoid the singular integrals when x is itself an element of the surface. But the problem is to get a constant independent of $\tau > 0$. Therefore we use the mappings from the surface to $\mathbb{R}^2$, translate differentiation with respect to x to differentiation with respect to y and apply integration by parts to transfer the differential operators to the layer function F. Furthermore we have to use a reduction result derived from a transformation formula for the Laplace operator. This helps us to get rid of higher order derivatives in direction y_3 which can not be treated with help of integration by parts, because we are only integrating over the variable y_1 and y_2. In the last section we prove that the results from Chapter 4 are applicable to geomathematics. We prove that the system of mass point representations as well as the systems of inner and outer harmonics are dense in $H^{m,2}(\partial\Sigma)$ for arbitrary $m \in \mathbb{N}$. With help of the results proved in Section 4.5 and Section 4.6, we are able to define $U_1^{\pm\tau}[F]$ for each $F \subset (H^{m,2}(\partial\Sigma))'$ and $\tau \in [0,\tau_0]$, where Σ is an outer C^{m+2}-domain. Moreover we are able

1.2. LIMIT FORMULAE AND JUMP RELATIONS OF POTENTIAL THEORY

to prove that the limit formula for U_1 even holds in this abstract setting. This enables us to extend the result from [FM03] and [FM04] about the density of the function systems in $L^2(\partial\Sigma)$.

In Chapter 2 we give the definitions of the surfaces and function spaces we will use, as well as some properties of them. Furthermore, we formulate results which will be important tools in the proofs of the main results. Chapter 4 is organized as follows. In Section 4.1 we define the layer potentials as well as their first order normal derivatives. These will be subject of our investigation in the following sections. Also some basic properties are stated. In Section 4.2 we state the limit formulae of potential theory for the first time, pointwisely and uniformly, i.e., in $C^0(\partial\Sigma)$-norm. These results will be the basis of our further considerations. They are mainly taken from literature, except of the limit formula for $\frac{\partial U_2}{\partial \nu}$. In Section 4.3 we present the first of our three main results. This is the convergence of the limit formulae in $C^m(\partial\Sigma)$-norm. The second one follows in Section 4.4 as the extension to the convergence even in the Hölder spaces $C^{m,\beta}(\partial\Sigma)$. Section 4.5 is essential for the proof of the convergence in $H^{m,2}(\partial\Sigma)$-norm, which is proved in Section 4.6. It contains the convergence in $L^2(\partial\Sigma)$, which is published in [Ker80] as well as in [Fre80] and has its origin in [Geh70]. Moreover, an important result about the continuity of the potential operators with respect to the $L^2(\Sigma)$-norm is stated and proved. Finally we prove the density of several function systems from geomathematics in $H^{m,2}(\partial\Sigma)$ as well as the limit formula of U_1 for $F \in (H^{m,2}(\partial\Sigma))'$ in Section 4.7.

The main progress achieved in Chapter 4 can be summarized by the following core results:

1. The convergence of the limit formula for $\frac{\partial U_2}{\partial \nu}$, pointwisely, in $C^0(\partial\Sigma)$-norm and in $L^2(\partial\Sigma)$-norm is proved, see Theorem 4.2.2.

2. The convergence of the limit formula for U_1, U_2, $\frac{\partial U_1}{\partial \nu}$ and $\frac{\partial U_2}{\partial \nu}$ in $C^m(\partial\Sigma)$-norm, $m \geq 1$ is proved, see Theorem 4.3.3.

3. Moreover this convergence is shown to be even a convergence in $C^{m,\beta}(\partial\Sigma)$-norm, $m \geq 0$ and $0 < \beta < 1$, see Theorem 4.4.1.

4. Finally, the convergence in $H^{m,2}(\partial\Sigma)$ with $F \in H^{m,2}(\partial\Sigma)$ for U_1, $F \in H^{m+1,2}(\partial\Sigma)$ for $\frac{\partial U_1}{\partial \nu}$ and U_2 and $F \in H^{m+2,2}(\partial\Sigma)$ for $\frac{\partial U_2}{\partial \nu}$ is proved, see Theorem 4.6.1.

5. An application of the limit formulae to geomathematics is given. We prove that the systems of mass distributions and outer harmonics, which are known to be dense in $L^2(\partial\Sigma)$, are even dense in $H^{m,2}(\partial\Sigma)$, $m \in \mathbb{N}$, for an outer $C^{2m-1,1}$-domain Σ, see Theorem

4.7.8. We also prove that the system of inner harmonics is dense in $H^{m,2}(\partial\Sigma)$. Moreover we extend U_1 to an operator on $\in (H^{m,2}(\partial\Sigma))'$ and we prove that the limit formula for U_1 even holds in this setting.

Outlook

In Chapter 4 we use the transformation formula of the Laplace operator for harmonic functions, in order to reduce second order normal derivatives of a given function F to a sum of first order normal derivatives of F, tangential derivatives of F as well as F itself. Using this technique, it is possible to reduce arbitrary normal derivatives of the potentials of the n-th layer on $\partial\Sigma$ to tangential derivatives of U_1, U_2 as well as of their first order normal derivatives, applied to tangential derivatives of F. Then all results contained in this dissertation might be translated to the limit formulae, and consequently the jump relations, for all remaining potentials, using the results and techniques presented. For details see also [Mül51], where the author uses this technique in order to get the jump relations for arbitrary derivatives of n-layer potentials. If we want to provide this results we are faced with two main problems. Of course, we have to identify the tangential differential operators at first. But then we also have to investigate the coefficient functions, because the operators are defined on the parallel surfaces $\partial\Sigma^{\pm\tau}$ and may depend on τ. Then also the convergence of these operators, when τ tends to zero, as well as the limiting operators, have to be investigated. An elementary approach has been done in the master thesis by M. Omari, cf. [Oma09], which I co-supervised. Another task might be the regularity assumptions in $H^{m,2}(\partial\Sigma)$ norm. When comparing the results to those in $C^m(\partial\Sigma)$ norm, we see that we have at least one order higher assumptions on F and Σ. It might be possible to achieve our techniques in such a way that we get assumptions in $H^{m,2}(\partial\Sigma)$ norm of the same strength as in $C^m(\partial\Sigma)$ norm. One possibility is to investigate the convergence of the the potentials

$$U[F](x) := \int_{\partial\Sigma} F(y) \frac{\partial}{\partial \underline{t}_i(y)} \frac{1}{|x-y|} \, dH^2(y),$$

$i = 1, 2$ and $x \in \mathbb{R}^3 \setminus \partial\Sigma$. Therefore results form [FM04, Paragraph 3.3.2] might be a starting point. The next point we want to mention concerns the jump relations. We were mainly interested in the limit formulae and took the jump relations as corollary. If one is interested only in the jump relations, there is the possibility to prove results under much weaker assumptions as for the limit formulae. The simple reason is that the singular integrals for $x \in \partial\Sigma$ do not occur in the jump relations. For such results see [FM04] where jump relations in $C^0(\partial\Sigma)$ norm

1.2. LIMIT FORMULAE AND JUMP RELATIONS OF POTENTIAL THEORY

are proved, [Mül51] where jump relations in $C^m(\partial\Sigma)$ norm are proved, or [Ker80] where jump relations in $L^2(\partial\Sigma)$ norm are proved. In Section 4.7 we prove that the system of mass point representations and the outer harmonics are dense in $H^{m,2}(\partial\Sigma)$. Application of these results to geomathematical problems is left to be implemented. In [FM03] or [FM04], where the authors prove these function systems to be dense in $L^2(\partial\Sigma)$, it is also proved that the normal derivatives of the function systems are dense in $L^2(\partial\Sigma)$. Density of the normal derivatives in $H^{m,2}(\partial\Sigma)$ seems also be provable from our point of view. Therefore it might be necessary to have $\frac{\partial U_1}{\partial \nu}, U_2 :$ $H^{m,2}(\partial\Sigma) \to H^{m,2}(\partial\Sigma)$. Moreover the density of oblique derivatives of the treated function systems in $H^{m,2}(\partial\Sigma)$ for $F \in H^{n,2}(\partial\Sigma)$ could be point of further research. It is left to investigate how far the results of this dissertation can be generalized to this case. Such a result would be an achievement for numerics of the oblique boundary problem for the Laplace equation, because until now density is only proved in the pre Hilbert space $\left(C^{0,\alpha}(\partial\Sigma), \|\cdot\|_{L^2(\partial\Sigma)}\right)$, see [FM04, Section 3.3.3]. Also an extension of our results to other function systems might be possible, c.f. [FM04]. In [FK80] the density of the system of mass point representations in $C^{0,\alpha}(\partial\Sigma)$ is proved. With help of the results from Chapter 4 it might be possible to extend this result to $C^{m,\beta}(\partial\Sigma)$. Another possible application are boundary problems for the Poisson equation, see e.g. [CK83] and [Gün57]. Here we would end up with some singular integral operators. At the end of this outlook we want to mention that a extension to the more general Lyapunov surfaces is possible, see e.g. [Gün57] or [Mic72]. Moreover a generalization to $\mathbb{R}^n$ might be possible, although it is not required from the view of geomathematical applications and there is only few literature.

Chapter 2

Preliminaries

This chapter contains the definitions and lemmata which will be used in the following chapters to derive the main results of this dissertation. In Chapter 2 we denote constants by $c_1, c_2, \ldots$, while we use $C_1, C_2, \ldots$ in Chapter 3 and Chapter 4. Because the constants in Chapter 4 will not depend on those from Chapter 3, we start again with $C_1, C_2, \ldots$ in Chapter 4. Vector valued functions, except of the normal vector field, are denoted by underlined letters and the euclidean scalar product of $x, y \in \mathbb{R}^n$ is denoted by $\langle x, y \rangle$ or (x, y), while $|x|$ is the euclidean norm. All functions are assumed to be real or real vector valued. The dual paring of $F \in H'$ and $G \in H$ is denoted by $_{H'}\langle F, G \rangle_H$ or $F(G)$, respectively. $\|G\|_H$ means the norm of G for a normed space $(H, \|\cdot\|_H)$.

2.1 Domains and Surfaces

We start with the definition of the surfaces and domains.

Definition 2.1.1. $\partial \Sigma \subset \mathbb{R}^n$ is called a $C^{m,\alpha}$-surface, $m \in \mathbb{N}$ and $0 \leq \alpha \leq 1$ and Σ is called a bounded $C^{m,\alpha}$-domain, if and only if

1. Σ is a bounded subset of $\mathbb{R}^n$ which is a domain, i.e., open and connected,

2. There exists an open cover $(U_i)_{i=1,\ldots,N}$ of $\partial \Sigma$ and corresponding $C^{m,\alpha}$-diffeomorhisms $\Psi_i : B_1^{\mathbb{R}^n}(0) \to U_i$, $i = 1, \ldots, N$, such that

$$\Psi_i : B_1^0(0) \to U_i \cap \partial \Sigma,$$
$$\Psi_i : B_1^+(0) \to U_i \cap \Sigma,$$

$$\Psi_i : B_1^-(0) \to U_i \cap \mathbb{R}^n \backslash \overline{\Sigma},$$

where $B_1^{\mathbb{R}^n}(0)$ denotes the open unit ball in $\mathbb{R}^n$, i.e., all $x \in \mathbb{R}^n$ with $|x| < 1$. $B_1^0(0)$ denotes the set of all $x \in B_1^{\mathbb{R}^n}(0)$ with $x_n = 0$, $B_1^+(0)$ denotes the set of all $x \in B_1^{\mathbb{R}^n}(0)$ with $x_n > 0$ and $B_1^-(0)$ denotes the set of all $x \in B_1^{\mathbb{R}^n}(0)$ with $x_n < 0$.

On the other hand Σ is called an *outer $C^{m,\alpha}$-domain*, if and only if $\Sigma \subset \mathbb{R}^n$ is open, connected and representable as $\Sigma := \mathbb{R}^n \backslash \overline{D}$, where D is a bounded $C^{m,\alpha}$-domain such that $0 \in D$. Consequently, $\partial \Sigma = \partial D$ is also a $C^{m,\alpha}$-surface. Ψ_i is called $C^{m,\alpha}$-diffeomorphism if and only if it is bijective, $(\Psi_i)_j \in C^{m,\alpha}\left(\overline{B_1^{\mathbb{R}^n}(0)}\right)$, $(\Psi_i^{-1})_j \in C^{m,\alpha}(\overline{U_i})$, $j = 1, 2, 3$ and we have for the determinant of the Jacobian Matrix of Ψ_i, $\text{Det}(D\Psi_i) \neq 0$ in $\overline{B_1^{\mathbb{R}^n}(0)}$. Furthermore, we find a C^∞-partition of $(w_i)_{1 \leq i \leq N}$ on $\partial \Sigma$ corresponding to the open cover $(U_i)_{1 \leq i \leq N}$, provided by [Alt02, Lemma 2.19].

For this definition and further details see e.g. [Dob06] or [GT01]. $\partial \Sigma$ is a compact double-pointfree $(n-1)$-dimensional $C^{m,\alpha}$-submanifold with $\partial(\partial \Sigma) = \emptyset$, i.e., $\partial \Sigma$ is a closed manifold. Let $\nabla_{\partial \Sigma}$ denote the gradient on $\partial \Sigma$ and $T(\partial \Sigma)$ the tangent space of $\partial \Sigma$. Because $\partial \Sigma$ is a submanifold of $\mathbb{R}^n$, we consider elements of $T(\partial \Sigma)$ as vectors in $\mathbb{R}^n$. Furthermore we can find orthonormal vector fields $\{\underline{t}_1, \ldots, \underline{t}_{n-1}\}$ on $\partial \Sigma$, generating $T(\partial \Sigma)$. As well $\underline{t}_i$, $1 \leq i \leq n-1$ as the outer unit normal vector ν are C^{m-1}-vector fields. H^{n-1} denotes the $(n-1)$-dimensional Hausdorff measure on $\partial \Sigma$, see [Alt04, Section 5.7], and λ^n the Lebesgue measure in $\mathbb{R}^n$. Throughout this dissertation we assume at least a Lipschitz boundaries, i.e., $C^{0,1}$-boundaries $\partial \Sigma$. Then we have $\nu \in L^\infty(\partial \Sigma; \mathbb{R}^n)$. For each differentiable function F, defined on $\partial \Sigma$, we have

$$\nabla_{\partial \Sigma} F := \sum_{i=1}^{n-1} \underline{t}_i \partial_{\underline{t}_i} F.$$

The definition is independent of the basis chosen. For details see e.g. [Alt04]. We have the following for $C^{m,\alpha}$-surfaces.

Lemma 2.1.2. Let $\partial \Sigma$ be a $C^{m,\alpha}$-surface, $m \in \mathbb{N}$, $m \in \mathbb{N}$, $\alpha \in [0,1]$. Then $\partial \Sigma$ is a $C^{n,\beta}$-surface, provided it holds $n \leq m$ and $n + \beta \leq m + \alpha$.

Proof. For the proof we refer to [CK83] or [GT01, Section 6.2]. □

In Figure 1, such a $C^{m,\alpha}$-surface is illustrated.

2.1. DOMAINS AND SURFACES

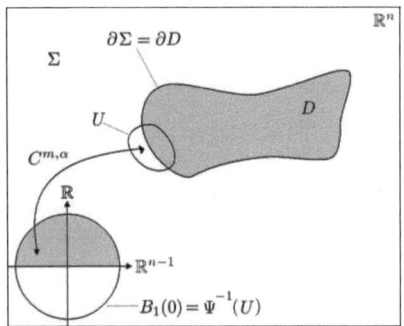

Figure 1: $C^{m,\alpha}$-surface

Note that some geomathematical relevant examples are even C^∞-surfaces, e.g. a sphere or an ellipsoid. At next, we define integration over $\partial\Sigma$.

Definition 2.1.3. The integral over a $C^{0,1}$-surface $\partial\Sigma$ of a function F is defined by:

$$\int_{\partial\Sigma} F(x)\, dH^{n-1}(x) := \sum_{j=1}^{N} \int_{\mathbb{R}^{n-1}} w_j(\Psi_j(x',0)) F(\Psi_j(x',0)) J_j(x')\, d\lambda^{n-1}(x').$$

The functions J_j are defined by

$$J_j^2(x') := \sum_{i=1}^{n} \left(J_j^k(x')\right)^2$$

for all $x' \in B_1^{\mathbb{R}^{n-1}}(0)$, $j = 1, \ldots, N$, where $J_j^k(x')$ is the determinant of the matrix $D\Psi_j(x',0)$ when suppressing the last column and the line with index k.

The integral is independent from the choice of $(U_i)_{1 \leq i \leq N}$, $(w_i)_{1 \leq i \leq N}$ and $(\Psi_i)_{1 \leq i \leq N}$. Furthermore we have the following result for the functions J_j.

Lemma 2.1.4. Let $\partial\Sigma$ be a $C^{m,\alpha}$-surface, $0 \leq \alpha \leq 1$ and $m \geq 1$. Then we have $J_j \in C^{m-1}(B_1^{\mathbb{R}^{n-1}}(0))$ for all $j \in \{1, \ldots, N\}$ with

$$0 < c_1^j \leq J_j(x') \leq c_2^j < \infty,$$

for all $x' \in B_1^{\mathbb{R}^{n-1}}(0)$, $j = 1, \ldots, N$.

Proof. This follows by the definition of J_j if we use the fact that $\text{Det}(D\Psi_j(x))$ is bounded on $\overline{B_1^{\mathbb{R}^n}(0)}$ and that $|\text{Det}(D\Psi_j(x))|$ is strictly positive for on $\overline{B_1^{\mathbb{R}^n}(0)}$ by Definition 2.1.1 and the expansion formula for determinants. For further details see e.g. [DL88, Appendix Chapter IV]. □

Boundaries of outer $C^{m,\alpha}$-domains in $\mathbb{R}^3$

In the last part of this subsection we give some further definitions and properties for outer $C^{m,\alpha}$-domains in $\mathbb{R}^3$. Therefore, let Σ be at least an outer C^2-domain in $\mathbb{R}^3$, if not stated otherwise. For such a domain we have the following.

Lemma 2.1.5. Let Σ be an outer C^2-domain. We find a constant $0 < \tau_0 < \infty$ such that the parallel surfaces $\partial\Sigma^\tau := \left\{ x + \tau\nu(x) \middle| x \in \partial\Sigma \right\}$ are well defined for all $\tau \in [-\tau_0, \tau_0]$ in the sense that for each $x \in \partial\Sigma^\tau$ there exists exactly one $y \in \partial\Sigma$ such that $x = y + \tau\nu(y)$. Additionally we have $\nu(x) = \nu(y)$ for $x \in \partial\Sigma$, where $y = x + \tau\nu(x) \in \partial\Sigma^\tau$ and ν denotes the normal vector field of $\partial\Sigma$.

Proof. We have to choose τ_0 such small that $1 - 2\tau_0 H + \tau_0^2 K$ remains positive on $\partial\Sigma$, where H is he mean curvature of $\partial\Sigma$ and K is the Gaussian curvature of $\partial\Sigma$. For details we refer to [CK83, Section 2.1] or [FM04, Section 3.1.1]. □

For the rest of this paragraph as well as in Chapter 4 we fix for each outer C^2-domain in $\mathbb{R}^3$, $0 < \tau_0 < \infty$ as a constant such that Lemma 2.1.5 holds and $x \pm \tau\nu(x) \notin \partial\Sigma$ for all $x \in \partial\Sigma$ and $\tau \in (0, \tau_0]$. We define

$$B_{\tau_0}(\partial\Sigma) := \bigcup_{\tau \in [-\tau_0, \tau_0]} \partial\Sigma^\tau.$$

At next, we define for each $C^{m,\alpha}$-surface $\partial\Sigma$ in $\mathbb{R}^3$ two tangential vectors as well as the normal vector forming an orthonormal basis of $\mathbb{R}^3$ at each point $x \in \partial\Sigma$. Before we close this section with the final lemma, we illustrate the set $B_{\tau_0}(\partial\Sigma)$ in the following figure.

2.1. DOMAINS AND SURFACES

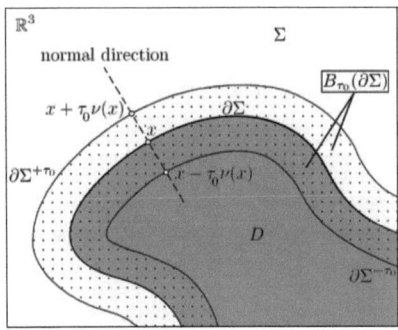

Figure 2: $B_{\tau_0}(\partial\Sigma)$ for an outer $C^{m,\alpha}$-domain

Lemma 2.1.6. Let $\partial\Sigma$ be a $C^{m,\alpha}$-surface, $m \in \mathbb{N}$, $m \geq 1$, $\alpha \in [0,1]$. We define

$$\underline{t}_1(x) := \frac{\sum_{i=1}^{N} w_i(x)(\partial_1 \Psi_i)(\Psi_i^{-1}(x))}{|\sum_{i=1}^{N} w_i(x)(\partial_1 \Psi_i)(\Psi_i^{-1}(x))|},$$

and

$$\underline{t}_2(x) := \frac{\sum_{i=1}^{N} w_i(x)(\partial_2 \Psi_i)(\Psi_i^{-1}(x)) - \underline{t}_1(x) \left\langle \underline{t}_1(x), \left(\sum_{i=1}^{N} w_i(x)(\partial_2 \Psi_i)(\Psi_i^{-1}(x))\right)\right\rangle}{|\sum_{i=1}^{N} w_i(x)(\partial_2 \Psi_i)(\Psi_i^{-1}(x)) - \underline{t}_1(x) \left\langle \underline{t}_1(x), \left(\sum_{i=1}^{N} w_i(x)(\partial_2 \Psi_i)(\Psi_i^{-1}(x))\right)\right\rangle|},$$

for all $x \in \partial\Sigma$. Then we set

$$\nu(x) := \frac{\underline{t}_1(x) \times \underline{t}_2(x)}{|\underline{t}_1(x) \times \underline{t}_2(x)|},$$

for all $x \in \partial\Sigma$. We have $\nu, \underline{t}_1, \underline{t}_2 \in C^{m-1}(\partial\Sigma)$. Finally, we define vector fields ν, $\underline{t}_1$ and $\underline{t}_2$ on $B_{\tau_0}(\partial\Sigma)$ by

$$\nu(x + \tau\nu(x)) := \nu(x),$$
$$\underline{t}_1(x + \tau\nu(x)) := \underline{t}_1(x),$$
$$\underline{t}_2(x + \tau\nu(x)) := \underline{t}_2(x),$$

for all $x \in \partial\Sigma$, $\tau \in [-\tau_0, \tau_0]$.

Proof. Note that for each $y \in B_{\tau_0}(\partial\Sigma)$ there exists one and only one $x \in \partial\Sigma$ and $\tau \in [-\tau_0, \tau_0]$ such that $y = x + \tau\nu(x)$. Consequently the definitions above are well defined. The regularity follows immediately by the definition of the vector fields together with the regularity of the mappings and Lemma 2.2.9. For $i = 1\ldots, N$ we can alternatively define $\nu(\Psi_i)$ on U_i by

$$\nu(\Psi_i(x,0)) = \frac{\partial_1 \Psi_i(x,0) \times \partial_1 \Psi_i(x,0)}{|\partial_1 \Psi_i(x,0) \times \partial_1 \Psi_i(x,0)|},$$

for all $x \in B_1^{\mathbb{R}^2}(0)$, because the mappings $\Psi_i(\,\cdot\,,0)$ are at least C^1-diffeomorphisms from $B_1^{\mathbb{R}^2}(0)$ to $U_i \cap \partial\Sigma$ and consequently $\partial_1 \Psi_i$ and $\partial_2 \Psi_i$ are as well elements of $T(\partial\Sigma)$ as linear independent.

$\square$

2.2 Function Spaces

We go on by introducing the function spaces which will be important in the following. We start with the classical spaces of smooth functions and compactly supported smooth functions.

Definition 2.2.1. Let Σ be a bounded or an outer $C^{0,1}$-domain. We define

$$\begin{aligned}
C^\infty(\Sigma) &:= \{F : \Sigma \to \mathbb{R} | \partial_1^{s_1} \ldots \partial_n^{s_n} F \text{ is continuous on } \Sigma \\
&\qquad \text{for all multi indices } (s_1,\ldots,s_n) \in \mathbb{N}^n\}, \\
C^\infty(\mathbb{R}^n) &:= \{F : \Sigma \to \mathbb{R} | \partial_1^{s_1} \ldots \partial_n^{s_n} F \text{ is continuous on } \mathbb{R}^n \\
&\qquad \text{for all multi indices } (s_1,\ldots,s_n) \in \mathbb{N}^n\}, \\
C_0^\infty(\mathbb{R}^n) &:= \{F \in C^\infty(\mathbb{R}^n) | \mathrm{supp}(F) \text{ is a compact subset of } \mathbb{R}^n\}, \\
C^\infty(\partial\Sigma) &:= \left\{ F\big|_{\partial\Sigma} \Big| F \in C^\infty(\mathbb{R}^n) \right\}.
\end{aligned}$$

Here $\mathrm{supp}(F)$ denotes the support of the function F, i.e., all $x \in \mathbb{R}^n$ for which $F(x) \neq 0$ and $F\big|_{\partial\Sigma}$ denotes the restriction of the function F to $\partial\Sigma$. Next we introduce Sobolev spaces on bounded and outer $C^{0,1}$-domains Σ as well as on $C^{m,1}$-surfaces $\partial\Sigma$.

Definition 2.2.2. Let Σ be a bounded $C^{0,1}$-domain and $r \in \mathbb{N}$. We define

$$H^{r,2}(\Sigma) := \{F : \Sigma \to \mathbb{R} | \partial_1^{\alpha_1} \cdots \partial_n^{\alpha_n} F \in L^2(\Sigma) \text{ for all } \alpha_1 + \ldots + \alpha_n \leq r\},$$

$$\|F\|_{H^{r,2}(\Sigma)} := \left(\sum_{|\alpha|=0}^{r} \sum_{i=1}^{N} \|\partial^\alpha F\|_{L^2(\Sigma)}^2 \right)^{\frac{1}{2}}.$$

2.2. FUNCTION SPACES

Let Σ be an outer $C^{0,1}$-domain and $\varrho_1, \varrho_2, \varrho_3$ be continuous, positive functions defined on $\overline{\Sigma}$. We define

$$L^2_{\varrho_1}(\Sigma) := \left\{ F : \Sigma \to \mathbb{R} | F \text{ is measurable with } \int_\Sigma F^2(x) \varrho_1^2(x) d\lambda^n(x) < \infty \right\},$$

$$H^{1,2}_{\varrho_1,\varrho_2}(\Sigma) := \left\{ F \in L^2_{\varrho_1}(\Sigma) | \partial_i F \in L^2_{\varrho_2}(\Sigma), 1 \leq i \leq n \right\},$$

$$H^{2,2}_{\varrho_1,\varrho_2,\varrho_3}(\Sigma) := \big\{ F \in L^2_{\varrho_1}(\Sigma) | \partial_i F \in L^2_{\varrho_2}(\Sigma)$$

$$\text{and } \partial_i \partial_j F \in L^2_{\varrho_3}(\Sigma), 1 \leq j, i \leq n \big\},$$

$$\|F\|_{L^2_{\varrho_1}(\Sigma)} := \left(\int_\Sigma F^2(x) \varrho_1^2(x) d\lambda^n(x) \right)^{\frac{1}{2}},$$

$$\|F\|_{H^{1,2}_{\varrho_1,\varrho_2}(\Sigma)} := \left(\|F\|^2_{L^2_{\varrho_1}(\Sigma)} + \sum_{i=1}^n \|\partial_i F\|^2_{L^2_{\varrho_2}(\Sigma)} \right)^{\frac{1}{2}},$$

$$\|F\|^2_{H^{2,2}_{\varrho_1,\varrho_2,\varrho_3}(\Sigma)} := \left(\|F\|^2_{L^2_{\varrho_1}(\Sigma)} + \sum_{i=1}^n \|\partial_i F\|^2_{L^2_{\varrho_2}(\Sigma)} + \sum_{i=1}^n \sum_{j=1}^n \|\partial_i \partial_j F\|^2_{L^2_{\varrho_3}(\Sigma)} \right)^{\frac{1}{2}}.$$

Let $\partial \Sigma$ be a $C^{0,1}$-surface and $(w_i)_{1 \leq i \leq N}$ be the C^∞-partition of unity of $\partial \Sigma$ corresponding to the open cover from Definition 2.1.1. For a function F defined on $\partial \Sigma$ we obtain a function $\theta_i F$ defined on $\mathbb{R}^{n-1}$ by:

$$\theta_i F(y) := \begin{cases} (w_i F)(\Psi_i(y,0)) & y \in B_1^{\mathbb{R}^{n-1}}(0), \\ 0 & \text{otherwise.} \end{cases}$$

Let now $\partial \Sigma$ be a $C^{m,1}$-surface, $m \in \mathbb{N}$. Furthermore let $s \in \mathbb{R}$, $r \in \mathbb{N}$, with $s < m+1$ and $0 \leq r \leq m$. Then we define

$$H^{s,2}(\partial \Sigma) := \left\{ F : \partial \Sigma \to \mathbb{R} | \theta_i F \in H^{s,2}(\mathbb{R}^{n-1}), 1 \leq i \leq N \right\},$$

$$H^{r,\infty}(\partial \Sigma) := \left\{ F : \Sigma \to \mathbb{R} | \theta_i F \in H^{r,\infty}(\mathbb{R}^{n-1}), 1 \leq i \leq N \right\},$$

$$\|F\|_{H^{s,2}(\partial \Sigma)} := \left(\sum_{i=1}^N \|\theta_i F\|^2_{H^{s,2}(\mathbb{R}^{n-1})} \right)^{\frac{1}{2}},$$

$$\|F\|_{H^{r,\infty}(\partial \Sigma)} := \max_{0 \leq s_1 + \ldots + s_{n-1} \leq r, 1 \leq i \leq N} \left\{ \text{ess sup}_{B_1^{\mathbb{R}^{n-1}}(0)} \left(|\partial_1^{s_1} \cdots \partial_{n-1}^{s_{n-1}} \theta_i F| \right) \right\},$$

where $H^{0,p}(\partial \Sigma)$ is identical with $L^p(\partial \Sigma)$, $p \in \{2, \infty\}$. The spaces $H^{s,2}(\partial \Sigma)$ and $H^{r,2}(\Sigma)$ are Hilbert spaces, while the spaces $H^{r,\infty}(\partial \Sigma)$ are Banach spaces with respect to the norms given above, see e.g. [Ada75], [DL88] or [Dob06].

The spaces $H^{s,2}(\mathbb{R}^{n-1})$ are defined via the Fourier transformation, see e.g. [SR91]. Differentiation in the definition above has to be understood in sense of weak differentiation. Furthermore we have the following for the weighted Sobolev and Lebesgue spaces.

Lemma 2.2.3. The spaces $H^{2,2}_{\varrho_1,\varrho_2,\varrho_3}(\Sigma)$, $H^{1,2}_{\varrho_1,\varrho_2}(\Sigma)$ and $L^2_{\varrho_1}(\Sigma)$ are Hilbert spaces with respect to the norms given in Definition 2.2.2.

Proof. Clearly all norms are induced by a scalar product. So the spaces are pre Hilbert spaces and it is left to show that they are complete. We start with $L^2_{\varrho_1}(\Sigma)$. Let $(f_n)_{n\in\mathbb{N}} \subset L^2_{\varrho_1}(\Sigma)$ be a Cauchy sequence in $L^2_{\varrho_1}(\Sigma)$. Then $(f_n\varrho_1)_{n\in\mathbb{N}}$ is a Cauchy sequence in $L^2(\Sigma)$. Because $L^2(\Sigma)$ is a Hilbert space, we find a $f \in L^2(\Sigma)$ such that $f_n\varrho_1 \to f$ in $L^2(\Sigma)$ for n tending to infinity. Then we have $\frac{f}{\varrho_1} \in L^2_{\varrho_1}(\Sigma)$ with $f_n \to \frac{f}{\varrho_1}$ in $L^2_{\varrho_1}(\Sigma)$ for n tending to infinity. Consequently $L^2_{\varrho_1}(\Sigma)$ is complete, i.e., a Hilbert spaces. Assume now that we have a Cauchy sequence $(g_n)_{n\in\mathbb{N}}$ in $H^{1,2}_{\varrho_1,\varrho_2}(\Sigma)$ and a Cauchy sequence $(h_n)_{n\in\mathbb{N}}$ in $H^{2,2}_{\varrho_1,\varrho_2,\varrho_3}(\Sigma)$. Then we have Cauchy sequences $(g_n)_{n\in\mathbb{N}}$, $(h_n)_{n\in\mathbb{N}}$ in $L^2_{\varrho_1}(\Sigma)$, $(\partial_i g_n)_{n\in\mathbb{N}}$, $(\partial_i h_n)_{n\in\mathbb{N}}$ in $L^2_{\varrho_2}(\Sigma)$ for $i=1,\ldots,n$ and $(\partial_j \partial_i h_n)_{n\in\mathbb{N}}$ in $L^2_{\varrho_3}(\Sigma)$ for $i,j=1,\ldots,n$. Consequently we find limiting functions $g,h \in L^2_{\varrho_1}(\Sigma)$, $g_i, h_i \in L^2_{\varrho_2}(\Sigma)$ and $h_{ji} \in L^2_{\varrho_3}(\Sigma)$ for each of this Cauchy sequences. All what is left to show is that $g_i = \partial_i g$, $h_i = \partial_i h$ and $h_{ji} = \partial_j \partial_i h$ in sense of weak differentiation. We treat the case for $H^{1,2}_{\varrho_1,\varrho_2}(\Sigma)$, $H^{2,2}_{\varrho_1,\varrho_2,\varrho_3}(\Sigma)$ can be done in the same way. Let $R_x := \Sigma \cap \left(\times_{i=1}^n]x_i, x_i+1[\right)$, where $x \in \mathbb{Z}^n$ is arbitrary. We have

$$\|F\|_{H^{1,2}(R_x)} \leq \min_{R_x}\{\varrho_1, \varrho_2\} \|F\|_{H^{1,2}_{\varrho_1,\varrho_2}(\Sigma)},$$

for all $F \in H^{1,2}_{\varrho_1,\varrho_2}(\Sigma)$. This yields that $(g_n)_{n\in\mathbb{N}}$ is a Cauchy sequence in $H^{1,2}(R_x)$ and we find a limiting function $\tilde{g}$ in the Hilbert space $H^{1,2}(R_x)$. On one side we have $g_n \to \tilde{g}$ and $\partial_i g_n \to \partial_i \tilde{g}$ in $L^2(R_x)$. On the other side $g_n \to G$ and $\partial_i g_n \to g_i$ in $L^2(R_x)$, because

$$\|F\|_{L^2(R_x)} \leq \min_{R_x}\{\varrho_1\} \|F\|_{L^2_{\varrho_1}(\Sigma)},$$
$$\|G\|_{L^2(R_x)} \leq \min_{R_x}\{\varrho_1\} \|G\|_{L^2_{\varrho_2}(\Sigma)},$$

for all $F \in L^2_{\varrho_1}(\Sigma)$ and $G \in L^2_{\varrho_2}(\Sigma)$. This yields $g_i = \partial_i g$ on R_x outside a set N_x of λ^n-measure 0. Finally $g_i = \partial_i g$ on Σ outside the set $N := \bigcup_{x\in\mathbb{Z}^n}(N_x \cup \partial R_x)$, which has λ^n-measure 0 and the proof is done. $\square$

Additionally, the product and chain rule of differentiation are also available for weakly differentiable functions.

Lemma 2.2.4. Let $U \subset \mathbb{R}^n$ be open. Let $F \in H^{m,p}(U)$, $G \in H^{m,q}(U)$, $\frac{1}{p} + \frac{1}{q} = 1$, $1 \leq p, q \leq \infty$ and $m \in \mathbb{N}$ be given. Then $F \cdot G \in H^{m,1}(U)$ and the weak derivatives can be computed by the product rule. Furthermore, let $\Psi : \tilde{U} \to U$ be a $C^{0,1}$-diffeomorphism and $H \in H^{1,p}(U)$ be given. Then $H \circ \Psi \in H^{1,p}(\tilde{U})$ and the weak derivatives can be computed using the chain

2.2. FUNCTION SPACES

rule. Furthermore the transformation formula for the integral holds for all $H \in L^p(U)$, i.e., $H \circ \Psi \in L^p(\tilde{U})$ and

$$\int_U H(x)d\lambda^n(x) = \int_{\tilde{U}} H(\Psi(y))|\mathrm{Det}(D\Psi)(y)|d\lambda^n(y).$$

Proof. This result is taken from [Alt02, Lemma 2.24 and Lemma 2.25]. □

For the definition of the spaces $H^{m,p}(U)$, $m \in \mathbb{N}$, $1 \leq p \leq \infty$, for arbitrary open sets $U \subset \mathbb{R}^n$, we refer to [Alt02]. Next we give an useful isomorphism between the spaces $H^{m,\infty}(U)$ and $C^{m-1,1}(U)$

Lemma 2.2.5. Let $U \subset \mathbb{R}^n$ be open and bounded. Then the embedding

$$\mathrm{Id}: C^{k,1}(\overline{U}) \to H^{k+1,\infty}(U),$$

is an isomorphism for each $k \in \mathbb{N}$, in the sense that each $F \in H^{k+1,\infty}(U)$ posses an unique representative in $C^{k,1}(\overline{U})$.

Proof. This result is taken from [Alt02, Lemma 8.5]. □

We proceed with a result about several equivalent norms on $L^2(\partial \Sigma)$.

Lemma 2.2.6. Let $\partial \Sigma$ be a $C^{0,1}$-surface and $L^2(\partial \Sigma)$ defined by Definition 2.2.2. Then we have that the three norms

$$\|F\|_{L^2(\partial\Sigma)} := \left(\sum_{i=1}^N \|\theta_i F\|^2_{L^2(\mathbb{R}^{n-1})} \right)^{\frac{1}{2}},$$

$$\|F\|^*_{L^2(\partial\Sigma)} := \left(\int_{\partial\Sigma} F^2(y)\, dH^{n-1}(y) \right)^{\frac{1}{2}},$$

$$\|F\|^{**}_{L^2(\partial\Sigma)} := \left(\sum_{i=1}^N \|F(\Psi_i)\|^2_{L^2(B_1^{\mathbb{R}^{n-1}}(0))} \right)^{\frac{1}{2}},$$

for all $F \in L^2(\partial\Sigma)$, are equivalent.

Proof. We have

$$\|F\|_{L^2(\partial\Sigma)} = \sum_{i=1}^N \|w_i^2(\Psi_i) F^2(\Psi_i)\|^2_{L^2(\mathbb{R}^{n-1})} \leq \sum_{i=1}^N c_1^i \|w_i(\Psi_i) F^2(\Psi_i) J_i\|^2_{L^2(\mathbb{R}^{n-1})}$$

$$\leq \max\{c_1^i\} \int_{\partial\Sigma} F^2(y)\, dH^{n-1}(y) = \|F\|_{L^2(\partial\Sigma)}^*,$$

and

$$\|F\|_{L^2(\partial\Sigma)}^* = \sum_{i=1}^N \|w_i(\Psi_i)F^2(\Psi_i)J_i\|_{L^2(\mathbb{R}^{n-1})}^2 \leq \sum_{i=1}^N c_2^i \|F^2(\Psi_i)\|_{L^2(B_1^{\mathbb{R}^{n-1}}(0))}^2 \leq \max\{c_2^i\}\|F\|_{L^2(\partial\Sigma)}^{**}$$

for all $F \in L^2(\partial\Sigma)$, where we used the constants from Lemma 2.1.4 and the fact that $0 \leq w_i \leq 1$ for $i = 1, \ldots, N$. It is left to proof that $\|F\|_{L^2(\partial\Sigma)}^{**} \leq c_3 \|F\|_{L^2(\partial\Sigma)}$, for a constant $0 < c_3 < \infty$ and all $F \in L^2(\partial\Sigma)$. We start to estimate

$$\left(\|F\|_{L^2(\partial\Sigma)}^{**}\right)^2 = \left(\sum_{i=1}^N \|F(\Psi_i)\|_{L^2(B_1^{\mathbb{R}^{n-1}}(0))}^2\right)^{\frac{1}{2}} = \sum_{i=1}^N \int_{B_1^{\mathbb{R}^{n-1}}(0)} F^2(\Psi_i(y,0)) d\lambda^{n-1}(y)$$

$$= \sum_{i=1}^N \int_{B_1^{\mathbb{R}^{n-1}}(0)} w_i(\Psi_i(y,0)) F^2(\Psi_i(y,0)) d\lambda^{n-1}(y)$$

$$+ \sum_{i=1}^N \int_{B_1^{\mathbb{R}^{n-1}}(0)} (1 - w_i(\Psi_i(y,0))) F^2(\Psi_i(y,0)) d\lambda^{n-1}(y)$$

$$= \sum_{i=1}^N \int_{B_1^{\mathbb{R}^{n-1}}(0)} w_i(\Psi_i(y,0)) F^2(\Psi_i(y,0)) d\lambda^{n-1}(y)$$

$$+ \sum_{i=1}^N \sum_{j=1, j\neq i}^N \int_{B_1^{\mathbb{R}^{n-1}}(0)} w_j(\Psi_i(y,0)) F^2(\Psi_i(y,0)) d\lambda^{n-1}(y).$$

We use the transformation $T_{ij} : \Psi_i^{-1}(U_i \cap U_j \cap \partial\Sigma) \to B_1^{\mathbb{R}^{n-1}}(0)$, defined by $T_{ij}(x) := \Psi_j^{-1}(\Psi_i(x))$ for all $x \in \Psi_i^{-1}(U_i \cap U_j \cap \partial\Sigma) \subset B_1^{\mathbb{R}^{n-1}}(0)$. By the chain rule for C^1-mappings as well as the product rule for the determinant we find

$$0 < \min_{\Psi_i^{-1}(U_i \cap U_j \cap \partial\Sigma)} |\text{Det}(DT_{ij})| \leq \max_{\Psi_i^{-1}(U_i \cap U_j \cap \partial\Sigma)} |\text{Det}(DT_{ij})| < \infty.$$

Note that all terms in the integral are positive. We get by using the transformation formula for the integral

$$\left(\|F\|_{L^2(\partial\Sigma)}^{**}\right)^2$$

$$= \sum_{i=1}^N \int_{B_1^{\mathbb{R}^{n-1}}(0)} w_i(\Psi_i(y,0)) F^2(\Psi_i(y,0)) d\lambda^{n-1}(y)$$

$$+ \sum_{i=1}^N \sum_{j=1, j\neq i}^N \int_{\Psi_i^{-1}(U_i \cap U_j \cap \partial\Sigma)} w_j(\Psi_i(y,0)) F^2(\Psi_i(y,0)) d\lambda^{n-1}(y)$$

2.2. FUNCTION SPACES

$$= \sum_{i=1}^{N} \int_{B_1^{\mathbb{R}^{n-1}}(0)} w_i(\Psi_i(y,0)) F^2(\Psi_i(y,0)) d\lambda^{n-1}(y)$$

$$+ \sum_{i=1}^{N} \sum_{j=1, j\neq i}^{N} \int_{\Psi_j^{-1}(U_i \cap U_j \cap \partial\Sigma)} w_j(\Psi_j(y,0)) F^2(\Psi_j(y,0)) |\text{Det}(DT_{ij})| d\lambda^{n-1}(y)$$

$$\leq \sum_{i=1}^{N} \int_{B_1^{\mathbb{R}^{n-1}}(0)} w_i(\Psi_i(y,0)) F^2(\Psi_i(y,0)) d\lambda^{n-1}(y)$$

$$+ \sum_{i=1}^{N} \max_{\Psi_i^{-1}(U_i \cap U_j \cap \partial\Sigma)} |\text{Det}(DT_{ij})| \sum_{j=1, j\neq i}^{N} \int_{B_1^{\mathbb{R}^{n-1}}(0)} w_j(\Psi_j(y,0)) F^2(\Psi_j(y,0)) d\lambda^{n-1}(y)$$

$$\leq \sum_{i=1}^{N} \Big(1 + \sum_{j=1, j\neq i}^{N} \max_{\Psi_j^{-1}(U_i \cap U_j \cap \partial\Sigma)} |\text{Det}(DT_{ji})|\Big)$$

$$\cdot \sum_{i=1}^{N} \int_{B_1^{\mathbb{R}^{n-1}}(0)} w_i(\Psi_i(y,0)) F^2(\Psi_i(y,0)) d\lambda^{n-1}(y)$$

Furthermore we divide each set $U_i \cap U_j \cap \partial\Sigma$ into two sets

$$V_{ij} := \{x \in U_i \cap U_j \cap \partial\Sigma \big| w_i(x) \leq w_j(x)\},$$
$$W_{ij} := \{x \in U_i \cap U_j \cap \partial\Sigma \big| w_i(x) \geq w_j(x)\}.$$

Starting like above, we have

$$\sum_{i=1}^{N} \int_{B_1^{\mathbb{R}^{n-1}}(0)} w_i(\Psi_i(y,0)) F^2(\Psi_i(y,0)) d\lambda^{n-1}(y)$$

$$= \sum_{i=1}^{N} \int_{B_1^{\mathbb{R}^{n-1}}(0)} w_i^2(\Psi_i(y,0)) F^2(\Psi_i(y,0)) d\lambda^{n-1}(y)$$

$$+ \sum_{i=1}^{N} \sum_{j=1, j\neq i}^{N} \int_{B_1^{\mathbb{R}^{n-1}}(0)} w_j(\Psi_i(y,0)) w_i(\Psi_i(y,0)) F^2(\Psi_i(y,0)) d\lambda^{n-1}(y)$$

$$= \sum_{i=1}^{N} \int_{B_1^{\mathbb{R}^{n-1}}(0)} w_i^2(\Psi_i(y,0)) F^2(\Psi_i(y,0)) d\lambda^{n-1}(y)$$

$$+ \sum_{i=1}^{N} \sum_{j=1, j\neq i}^{N} \Big(\int_{\Psi_i^{-1}(V_{ij})} w_j(\Psi_i(y,0)) w_i(\Psi_i(y,0)) F^2(\Psi_i(y,0)) d\lambda^{n-1}(y)$$

$$+ \int_{\Psi_i^{-1}(W_{ij})} w_j(\Psi_i(y,0)) w_i(\Psi_i(y,0)) F^2(\Psi_i(y,0)) d\lambda^{n-1}(y) \Big)$$

$$\leq \sum_{i=1}^{N} \int_{B_1^{\mathbb{R}^{n-1}}(0)} w_i^2(\Psi_i(y,0)) F^2(\Psi_i(y,0)) d\lambda^{n-1}(y)$$

$$+ \sum_{i=1}^{N} \sum_{j=1, j\neq i}^{N} \Big(\int_{\Psi_j^{-1}(V_{ij})} w_j(\Psi_j(y,0)) w_i(\Psi_j(y,0)) F^2(\Psi_j(y,0)) |\mathrm{Det}(DT_{ij})| d\lambda^{n-1}(y)$$
$$+ \int_{B_1^{\mathbb{R}^{n-1}}(0)} w_i^2(\Psi_i(y,0)) F^2(\Psi_i(y,0)) d\lambda^{n-1}(y) \Big)$$
$$\leq \sum_{i=1}^{N} \int_{B_1^{\mathbb{R}^{n-1}}(0)} w_i^2(\Psi_i(y,0)) F^2(\Psi_i(y,0)) d\lambda^{n-1}(y)$$
$$+ \sum_{i=1}^{N} \sum_{j=1, j\neq i}^{N} \Big(\max_{\Psi_j^{-1}(U_i \cap U_j \cap \partial \Sigma)} |\mathrm{Det}(DT_{ji})| \int_{B_1^{\mathbb{R}^{n-1}}(0)} w_j^2(\Psi_j(y,0)) F^2(\Psi_j(y,0)) d\lambda^{n-1}(y)$$
$$+ \int_{B_1^{\mathbb{R}^{n-1}}(0)} w_i^2(\Psi_i(y,0)) F^2(\Psi_i(y,0)) d\lambda^{n-1}(y) \Big).$$

Finally we find that

$$c_3 = \Big(\sum_{i=1}^{N} \Big(1 + \sum_{j=1, j\neq 1}^{N} \max_{\Psi_j^{-1}(U_i \cap U_j \cap \partial \Sigma)} |\mathrm{Det}(DT_{ji})| \Big) \Big)$$
$$\cdot \Big(\sum_{i=1}^{N} \Big(N + \sum_{j=1, j\neq 1}^{N} \max_{\Psi_j^{-1}(U_i \cap U_j \cap \partial \Sigma)} |\mathrm{Det}(DT_{ji})| \Big) \Big),$$

is a possible choice. □

Whenever we use an equivalent norm, we neglect the equivalence constants to simplify the computations and avoid confusion. Before we present some important features of Sobolev functions defined on submanifolds we close this section with the following remark.

Remark 2.2.7. (i) The definition of the spaces on $\partial\Sigma$ above is independent from the choice of $(U_i)_{1\leq i\leq N}$, $(w_i)_{1\leq i\leq N}$ and $(\Psi_i)_{1\leq i\leq N}$. Moreover $C^\infty(\Sigma)$ is dense in $H^{r,2}(\Sigma)$ for all $r \in \mathbb{N}$ if Σ is assumed to be a bounded $C^{0,1}$-domain. If Σ is an outer $C^{0,1}$-domain we have that $C_0^\infty(\mathbb{R}^n)\big|_\Sigma$ is dense in $H^{m,2}(\Sigma)$. Moreover, if $\partial\Sigma$ is a C^{m+1}-surface, then we have that $C^\infty(\partial\Sigma)$ is dense in $H^{s,2}(\partial\Sigma)$ for all $s \in \mathbb{R}$, $s < m+1$, see [Ada75]. e.g. [Fic48]. For more details about Sobolev spaces see e.g. [Ada75], [DL88] or [Dob06].

(ii) There are many ways to introduce the spaces $H^{s,2}(\partial\Sigma)$ on a $C^{m,1}$-surface $\partial\Sigma$, $s < m+1$. Some examples for equivalent norms on $L^2(\partial\Sigma)$ are given in Lemma 2.2.6. More details can be found in e.g. [DL88]. In this dissertation we will use the following scalar product on $L^2(\partial\Sigma)$

2.2. FUNCTION SPACES

given by
$$\langle F, G \rangle_{L^2(\partial\Sigma)} := \int_{\partial\Sigma} F(y) \cdot G(y) \, dH^{n-1}(y),$$
for $F, G \in L^2(\partial\Sigma)$. We introduce the spaces $H^{-s,2}(\partial\Sigma)$ on a $C^{m,1}$-surface $\partial\Sigma$, $0 \leq s < m+1$, as follows. Identify each function $F \in L^2(\partial\Sigma)$ with a linear continuous functional on $H^{s,2}(\partial\Sigma)$, defined by
$$F(G) := \int_{\partial\Sigma} F(x) \cdot G(x) \, dH^{n-1}(x),$$
for all $G \in H^{s,2}(\partial\Sigma)$. Then $(H^{s,2}(\partial\Sigma))'$ is defined as
$$\left(H^{s,2}(\partial\Sigma)\right)' := \overline{L^2(\partial\Sigma)}\bigg|_{\|\cdot\|_{(H^{s,2}(\partial\Sigma))'}},$$
where
$$\|F\|_{(H^{s,2}(\partial\Sigma))'} := \sup_{G \in H^{s,2}(\partial\Sigma)} \frac{|F(G)|}{\|G\|_{H^{s,2}(\partial\Sigma)}}.$$
In this way we end up with the space $H^{-s,2}(\partial\Sigma)$ defined in the previous definition, see [Dob06]. Therefore we will keep the notation $H^{-s,2}(\partial\Sigma)$. We get the following chain of rigged Hilbert spaces, called Gelfand triple.
$$H^{s,2}(\partial\Sigma) \subset L^2(\partial\Sigma) \subset H^{-s,2}(\partial\Sigma),$$
densely and continuously. Additionally we have for the duality product
$$_{H^{-s,2}(\partial\Sigma)}\langle F, G \rangle_{H^{s,2}(\partial\Sigma)} = \int_{\partial\Sigma} F(x) \cdot G(x) \, dH^{n-1}(x),$$
for all $F \in L^2(\partial\Sigma)$. Analogously, we introduce $(H^{1,2}(\Sigma))'$ if Σ is a bounded $C^{0,1}$-domain as well as
$\left(H^{1,2}_{|x|^2,|x|^3}(\Sigma)\right)'$ if Σ is an outer $C^{0,1}$-domain. Then we get the Gelfand triples given in the next part of this remark.

(iii) In Chapter 3 we will use three different Gelfand triples. Namely, this are
$$H^{1,2}(\Sigma) \subset L^2(\Sigma) \subset \left(H^{1,2}(\Sigma)\right)',$$
for bounded $C^{0,1}$-domains,
$$H^{1,2}_{|x|^2,|x|^3}(\Sigma) \subset L^2_{|x|^2}(\Sigma) \subset \left(H^{1,2}_{|x|^2,|x|^3}(\Sigma)\right)',$$

for outer $C^{0,1}$-domains and
$$H^{\frac{1}{2},2}(\partial\Sigma) \subset L^2(\partial\Sigma) \subset H^{-\frac{1}{2},2}(\partial\Sigma),$$
for $C^{0,1}$-surfaces.

(iv) For a bounded $C^{0,1}$-domain Σ and $F \in H^{1,2}(\Sigma)$, $\operatorname{tr}(u)$ denotes the trace of F on $\partial\Sigma$. Here $\operatorname{tr} : H^{1,2}(\Sigma) \to L^2(\partial\Sigma)$ is the unique continuous operator having the property
$$\operatorname{tr}(F) = F|_{\partial\Sigma} \quad \text{for all } F \in C^0(\overline{\Sigma}) \cap H^{1,2}(\Sigma),$$
see [Alt02, Section A.6.6], where $\overline{\Sigma}$ means the closure of Σ. The statement also holds if Σ is an outer $C^{0,1}$-domain and if we replace $H^{1,2}(\Sigma)$ by $H^{1,2}_{\varrho_1,\varrho_2}(\Sigma)$.

Hölder spaces on subsets of $\mathbb{R}^3$

We close this section with the definition of the function spaces of Hölder continuous differentiable functions defined on open subsets $U \subset \mathbb{R}^3$ as well as on surfaces $\partial\Sigma \subset \mathbb{R}^3$. This spaces will be important in Chapter 4.

Definition 2.2.8. Let $U \subset \mathbb{R}^3$ open. For $n \in \mathbb{N}$ and $\beta \in [0,1]$ we define the spaces

$$C^0(U) := \left\{ f : U \to \mathbb{R} \,\middle|\, f \text{ is continuous on } U \right\},$$

$$C^{0,\beta}(U) := \left\{ f \in C^0(U) \,\middle|\, \sup\left\{ \frac{|f(x)-f(y)|}{|x-y|^\beta} \,\middle|\, x,y \in U, x \neq y \right\} < \infty \right\},$$

$$C^n(U) := \left\{ f : U \to \mathbb{R} \,\middle|\, \partial_1^{\gamma_1}\partial_2^{\gamma_2}\partial_3^{\gamma_3} f \in C^0(U) \text{ for all } \gamma \text{ with } \gamma_1+\gamma_2+\gamma_3 \leq n \right\},$$

$$C^{n,\beta}(U) := \left\{ f \in C^n(U) \,\middle|\, \partial_1^{\gamma_1}\partial_2^{\gamma_2}\partial_3^{\gamma_3} f \in C^{0,\beta}(U) \text{ for all } \gamma \text{ with } \gamma_1+\gamma_2+\gamma_3 = n \right\},$$

$$C^0(\overline{U}) := \left\{ f : \overline{U} \to \mathbb{R} \,\middle|\, f \text{ is continuous on } \overline{U} \right\},$$

$$C^{0,\beta}(\overline{U}) := \left\{ f \in C^0(\overline{U}) \,\middle|\, \sup\left\{ \frac{|f(x)-f(y)|}{|x-y|^\beta} \,\middle|\, x,y \in \overline{U}, x \neq y \right\} < \infty \right\},$$

$$C^n(\overline{U}) := \left\{ f : \overline{U} \to \mathbb{R} \,\middle|\, \partial_1^{\gamma_1}\partial_2^{\gamma_2}\partial_3^{\gamma_3} f \in C^0(\overline{U}) \text{ for all } \gamma \text{ with } \gamma_1+\gamma_2+\gamma_3 \leq n \right\},$$

$$C^{n,\beta}(\overline{U}) := \left\{ f \in C^n(\overline{U}) \,\middle|\, \partial_1^{\gamma_1}\partial_2^{\gamma_2}\partial_3^{\gamma_3} f \in C^{0,\beta}(\overline{U}) \text{ for all } \gamma \text{ with } \gamma_1+\gamma_2+\gamma_3 = n \right\},$$

$$C^\infty(U) := \left\{ f : U \to \mathbb{R} \,\middle|\, \partial_1^{\gamma_1}\partial_2^{\gamma_2}\partial_3^{\gamma_3} f \in C^{0,\beta}(U) \text{ for all } \gamma \in \mathbb{N}^3 \right\}.$$

2.2. FUNCTION SPACES

The spaces defined on U are normed vector spaces, while the spaces defined on $\overline{U}$ are Banach spaces, see [Alt02, Section 1.6], with the norms

$$\|f\|_{C^0(U)} := \sup\left\{|f(x)|\,\Big|\,x \in U\right\},$$

$$\|f\|_{C^{0,\beta}(U)} := \|f\|_{C^0(U)} + \sup\left\{\frac{|f(x)-f(y)|}{|x-y|^\beta}\,\Big|\,x,y \in U, x \neq y\right\},$$

$$\|f\|_{C^n(U)} := \sum_{\gamma_1+\gamma_2+\gamma_3=0}^{n} \|\partial_1^{\gamma_1}\partial_2^{\gamma_2}\partial_3^{\gamma_3} f\|_{C^0(U)},$$

$$\|f\|_{C^{n,\beta}(U)} := \|f\|_{C^{n-1}(U)} + \sum_{\gamma_1+\gamma_2+\gamma_3=n} \|\partial_1^{\gamma_1}\partial_2^{\gamma_2}\partial_3^{\gamma_3} f\|_{C^{0,\beta}(U)},$$

$$\|f\|_{C^0(\overline{U})} := \sup\left\{|f(x)|\,\Big|\,x \in \overline{U}\right\},$$

$$\|f\|_{C^{0,\beta}(\overline{U})} := \|f\|_{C^0(\overline{U})} + \sup\left\{\frac{|f(x)-f(y)|}{|x-y|^\beta}\,\Big|\,x,y \in \overline{U}, x \neq y\right\},$$

$$\|f\|_{C^n(\overline{U})} := \sum_{\gamma_1+\gamma_2+\gamma_3=0}^{n} \|\partial_1^{\gamma_1}\partial_2^{\gamma_2}\partial_3^{\gamma_3} f\|_{C^0(\overline{U})},$$

$$\|f\|_{C^{n,\beta}(\overline{U})} := \|f\|_{C^{n-1}(\overline{U})} + \sum_{\gamma_1+\gamma_2+\gamma_3=n} \|\partial_1^{\gamma_1}\partial_2^{\gamma_2}\partial_3^{\gamma_3} f\|_{C^{0,\beta}(\overline{U})}.$$

Note that $C^{n,0}(U)$ is not $C^n(U)$. Now, let $\partial\Sigma$ be a $C^{m,\alpha}$-surface, $m,n \in \mathbb{N}$, $n \leq m$, $\beta,\alpha \in [0,1]$ and $n + \beta \leq m + \alpha$. Then we define the spaces

$$C^{n,\beta}(\partial\Sigma) := \left\{f : \Sigma \to \mathbb{R}\,\Big|\,f(\Psi_i(\cdot,\cdot,0)) \in C^{n,\beta}(B_1^{\mathbb{R}^2}(0)), i=1,\ldots,N\right\}.$$

The spaces $C^{m,\alpha}(\partial\Sigma)$ are Banach spaces, see [GT01, Section 6.2], when equipped with the norms

$$\|f\|_{C^{n,\beta}(\partial\Sigma)} := \sum_{i=1}^{N} \|f(\Psi_i(\cdot,\cdot,0))\|_{C^{n,\beta}(B_1^{\mathbb{R}^2}(0))},$$

These definition is independent of $(\Psi_i)_{1 \leq i \leq N}$, $(U_i)_{1 \leq i \leq N}$ and $(w_i)_{1 \leq i \leq N}$. Furthermore $C^\infty(\partial\Sigma) \subset C^{n,\beta}(\partial\Sigma)$ for all $n+\beta \leq m+\alpha$. Consequently $C^{n,\beta}(\partial\Sigma) \cap H^{s,2}(\partial\Sigma)$ is dense in $H^{s,2}(\partial\Sigma)$ for $s < m$.

We have the following result about the spaces $C^{m,\alpha}(\partial\Sigma)$.

Lemma 2.2.9. Let $\partial\Sigma$ be a $C^{m,\alpha}$-surface, $m \in \mathbb{N}$, $\alpha \in [0,1]$. Furthermore, let U and V be sets and $\Psi : U \to V$ be a C^1-diffeomorphism. Assume we have $f, g \in C^{m,\alpha}(\partial\Sigma)$. Then $f + g, f \cdot g \in C^{m,\alpha}(\partial\Sigma)$ and $f \in C^{n,\beta}(\partial\Sigma)$ for all $n \in \mathbb{N}$, $n \leq m$, and $\beta \in [0,1]$, such that $n + \beta \leq n + \alpha$. Furthermore we have

$$\|f+g\|_{C^{m,\alpha}(\partial\Sigma)} \leq c_4^1 \left(\|f\|_{C^{m,\alpha}(\partial\Sigma)} + \|g\|_{C^{m,\alpha}(\partial\Sigma)} \right),$$
$$\|f \cdot g\|_{C^{m,\alpha}(\partial\Sigma)} \leq c_4^2 \|f\|_{C^{m,\alpha}(\partial\Sigma)} \cdot \|g\|_{C^{m,\alpha}(\partial\Sigma)},$$
$$\|f\|_{C^{n,\beta}(\partial\Sigma)} \leq c_4^3 \|f\|_{C^{m,\alpha}(\partial\Sigma)},$$
$$\|f(\Psi)\|_{C^{n,\beta}(V)} \leq c_4^4 \|f\|_{C^{n,\alpha}(U)}.$$

for a constant $0 < c_4^i < \infty$, $i = 1, 2, 3$, depending on f and g.

Proof. For $f \in C^{m,\alpha}(\partial\Sigma)$ we have $f \in C^{n,\beta}(\partial\Sigma)$ for $n < m$ and $\beta \in [0,1]$, see [Gün57, Section 2]. Finally,

$$\frac{|f(x) - f(y)|}{|x-y|^\beta} = |x-y|^{\alpha-\beta} \cdot \frac{|f(x) - f(y)|}{|x-y|^\beta},$$

for all $x, y \in B_1^{\mathbb{R}^2}(0)$, proves the last statement of the lemma. We use the following estimates in order to estimate the Hölder constants for $f + g$ and $f \cdot g$

$$\frac{|f(x) + g(x) - f(x) - g(x)|}{|x-y|^\alpha} \leq \frac{|f(x) - f(x)|}{|x-y|^\alpha} + \frac{|g(x) - g(x)|}{|x-y|^\alpha},$$
$$\frac{|f(x) \cdot g(x) - f(y) \cdot g(y)|}{|x-y|^\alpha} \leq f(x) \cdot \frac{|g(x) - g(y)|}{|x-y|^\alpha} + g(y) \cdot \frac{|f(x) - f(y)|}{|x-y|^\alpha}.$$

for all $x, y \in B_1^{\mathbb{R}^2}(0)$. Using the linearity of the differentiation, the product rule as well as the last statement of this lemma, also this part of the proof is done. The last inequality follows directly if we recall that each C^1-diffeomorphism is also a $C^{0,1}$-diffeomorphism. So we are able to find an $0 < c_4^i < \infty$ such that the lemma holds. Details can also be found in [GT01, Section 6.2] and [Gün57, Sections 2 and 18]. □

We want to mention, that continuous functions on compact sets as well as Hölder continuous functions on open sets U are equicontinuous, see e.g. [Gün57, Section 2]. For such functions we find unique continuations onto $\overline{U}$. The final lemma of this subsection states this existence of a unique continuation, which is even a Hölder continuous function up to the boundary.

Lemma 2.2.10. Let $\partial\Sigma$ be a regular $C^{m,\alpha}$-surface, $m, n \in \mathbb{N}_0$, $n \leq m$ and $0 \leq \alpha, \beta \leq 1$ such that $n + \beta \leq m + \alpha$. Furthermore let $f \in C^{n,\beta}(D)$ and $g \in C^{n,\beta}(\Sigma)$ be given. Then there exist continuations $\tilde{f} \in C^{n,\beta}(D \cup B_{r_0}(\Sigma))$ and $\tilde{g} \in C^{n,\beta}(\Sigma \cup B_{r_0}(\Sigma))$. Furthermore, $\tilde{f}$ and $\tilde{g}$ are uniquely determined on $\overline{D}$, or $\overline{\Sigma}$ respectively, as the unique continuous continuations of f and g.

Proof. These results can be found in [Gün57, Sections 2 and 18] and [GT01, Section 6.2]. □

2.3 Properties of Sobolev Spaces on Submanifolds

At next we give some useful lemmata about the spaces defined in the previous section.

Lemma 2.3.1. Let $\partial\Sigma$ be a $C^{0,1}$-surface. One has that

$$\nabla_{\partial\Sigma} : H^{\frac{1}{2},2}(\partial\Sigma) \to H^{-\frac{1}{2},2}(\partial\Sigma; T(\partial\Sigma)),$$

is continuous, i.e., there exists $0 < c_5 < \infty$ such that

$$\|\nabla_{\partial\Sigma} F\|_{H^{-\frac{1}{2},2}(\partial\Sigma;T(\partial\Sigma))} \leq c_5 \|F\|_{H^{\frac{1}{2},2}(\partial\Sigma)}$$

for all $F \in H^{\frac{1}{2},2}(\partial\Sigma)$.

Proof. For the proof see [GR06, Lemma 2.2.4]. □

For inner domains we have a Poincaré inequality, which is missing for outer domains.

Lemma 2.3.2. For all $F \in H^{1,2}(\Sigma)$ the following inequality is valid

$$\int_\Sigma \langle \nabla F, \nabla F \rangle \, d\lambda^n + \int_{\partial\Sigma} F^2 \, dH^{n-1} \geq c_6 \left(\int_\Sigma F^2 \, d\lambda^n + \int_\Sigma \langle \nabla F, \nabla F \rangle \, d\lambda^n \right),$$

for a constant $0 < c_6 < \infty$.

Proof. For the proof see [GR06, Lemma 2.7]. □

We go on with a lemma about the traces on the boundary of functions from Sobolev spaces defined in inner or outer domains.

Lemma 2.3.3. Let Σ be a bounded or on outer $C^{0,1}$-domain. For all $F \in H^{1,2}(\Sigma)$ one has $\operatorname{tr}(F) \in H^{\frac{1}{2},2}(\partial\Sigma)$ with

$$\|\operatorname{tr}(F)\|_{H^{\frac{1}{2},2}(\partial\Sigma)} \leq c_7 \|F\|_{H^{1,2}(\Sigma)},$$

where $0 < c_7 < \infty$. Conversely, for all $F \in H^{\frac{1}{2},2}(\partial\Sigma)$ there exists a $\tilde{F} \in H^{1,2}(\Sigma)$ such that $\operatorname{tr}(\tilde{F}) = F$ and

$$\|\tilde{F}\|_{H^{1,2}(\Sigma)} \leq c_8 \|F\|_{H^{\frac{1}{2},2}(\partial\Sigma)},$$

where $0 < c_8 < \infty$.

Proof. For the proof see [GR06, Lemma 2.2.5]. □

We can prove an analogon to Lemma 2.3.3 for weighted Sobolev spaces introduced in Definition 2.2.2.

Lemma 2.3.4. Let Σ be an outer $C^{0,1}$-domain and $\varrho_1, \varrho_2 > 0$ continuous functions on $\overline{\Sigma}$. For all $F \in H^{1,2}_{\varrho_1,\varrho_2}(\Sigma)$ one has $\operatorname{tr}(F) \in H^{\frac{1}{2},2}(\partial\Sigma)$ with

$$\|\operatorname{tr}(F)\|_{H^{\frac{1}{2},2}(\partial\Sigma)} \leq c_{10} \|F\|_{H^{1,2}_{\varrho_1,\varrho_2}(\Sigma)},$$

where $0 < c_{10} < \infty$.

Proof. For simplicity we write F instead of $\operatorname{tr}(F)$ in the following. So we have to show that

$$\|F\|_{H^{\frac{1}{2},2}(\partial\Sigma)} \leq c_{10} \|F\|_{H^{1,2}_{\varrho_1,\varrho_2}(\Sigma)},$$

with a constant $0 < c_{10} < \infty$. Assume $0 < R < \infty$ such that $\partial\Sigma \subset B_R(0)$ and choose a $w \in C_0^\infty(B_{2R}(0))$ with $w = 1$ on $B_R(0)$. Such a w exists, see [Alt02]. We obviously have $Fw = F$ on $\partial\Sigma$. Additionally $Fw \in H^{1,2}(\Sigma)$ and consequently, using Lemma 2.3.3, we can estimate

$$\|F\|_{H^{\frac{1}{2},2}(\partial\Sigma)} = \|Fw\|_{H^{\frac{1}{2},2}(\partial\Sigma)} \leq \|Fw\|_{H^{1,2}(\Sigma)}.$$

Because ϱ_1 and ϱ_2 are strictly positive on $\overline{\Sigma \cap B_{2R}(0)}$ we have that $\frac{1}{\varrho_1}$ and $\frac{1}{\varrho_2}$ are bounded on this compact set. So, we can finish the proof by estimating

$$\|F\|_{H^{\frac{1}{2},2}(\partial\Sigma)} \leq c_{10} \|F\|_{H^{1,2}_{\varrho_1,\varrho_2}(\Sigma)},$$

where c_{10} is the maximum of the continuous functions ϱ_1 and ϱ_2 on the compact set $\overline{\Sigma \cap B_{2R}(0)}$. □

2.3. PROPERTIES OF SOBOLEV SPACES ON SUBMANIFOLDS

The next lemma states an useful estimate of the dual paring.

Lemma 2.3.5. Let Σ be a bounded $C^{0,1}$-domain. We have

$$\left| {}_{H^{1,2}(\Sigma)}\langle F, G\rangle_{(H^{1,2}(\Sigma))'} \right| \leq \|F\|_{H^{1,2}(\Sigma)} \|G\|_{(H^{1,2}(\Sigma))'},$$

for all $F \in H^{1,2}(\Sigma)$ and all $G \in (H^{1,2}(\Sigma))'$. Let $\partial\Sigma$ be a $C^{0,1}$-surface. We have

$$\left| {}_{H^{\frac{1}{2},2}(\Sigma)}\langle F, G\rangle_{H^{-\frac{1}{2},2}(\Sigma)} \right| \leq \|F\|_{H^{\frac{1}{2},2}(\Sigma)} \|G\|_{H^{-\frac{1}{2},2}(\Sigma)},$$

for all $F \in H^{\frac{1}{2},2}(\Sigma)$ and all $G \in H^{-\frac{1}{2},2}(\Sigma)$. Let Σ be a outer $C^{0,1}$-domain. We have

$$\left| {}_{H^{1,2}_{\varrho_1,\varrho_2}(\Sigma)}\langle F, G\rangle_{\left(H^{1,2}_{\varrho_1,\varrho_2}(\Sigma)\right)'} \right| \leq \|F\|_{H^{1,2}_{\varrho_1,\varrho_2}(\Sigma)} \|G\|_{\left(H^{1,2}_{\varrho_1,\varrho_2}(\Sigma)\right)'},$$

for all $F \in H^{1,2}_{\varrho_1,\varrho_2}(\Sigma)$ and all $G \in \left(H^{1,2}_{\varrho_1,\varrho_2}(\Sigma)\right)'$.

Proof. We have the dense embeddings of the function spaces into the respective dual spaces. Consequently [BK95, Section 1.1] gives the claimed inequalities. □

We proceed with a result about multiplier functions in $H^{\frac{1}{2},2}(\partial\Sigma)$.

Lemma 2.3.6. Let $\partial\Sigma$ be a $C^{0,1}$-surface, $F \in H^{\frac{1}{2},2}(\partial\Sigma)$ and $G \in H^{1,\infty}(\partial\Sigma)$. Then we have for a constant $0 < c_9 < \infty$

$$\|GF\|_{H^{\frac{1}{2},2}(\partial\Sigma)} \leq c_9 \|F\|_{H^{\frac{1}{2},2}(\partial\Sigma)}.$$

Proof. For the proof see [GR06, Lemma 2.2.6]. □

Next we prove a result analogous to Lemma 2.3.6 for distributions in $H^{-\frac{1}{2},2}(\partial\Sigma)$.

Lemma 2.3.7. Let $\partial\Sigma$ be a $C^{0,1}$-surface, $F \in H^{-\frac{1}{2},2}(\partial\Sigma)$ and $G \in H^{1,\infty}(\partial\Sigma)$. Then we have $FG \in H^{-\frac{1}{2},2}(\partial\Sigma)$ with

$$\|FG\|_{H^{-\frac{1}{2},2}(\partial\Sigma)} \leq c_9 \|G\|_{H^{-\frac{1}{2},2}(\partial\Sigma)},$$

where $0 < c_9 < \infty$ is the constant from Lemma 2.3.6 and FG is defined by

$$FG(H) := F(GH),$$

for all $H \in H^{\frac{1}{2},2}(\partial\Sigma)$.

Proof. Since $GH \in H^{\frac{1}{2},2}(\partial\Sigma)$ by Lemma 2.3.6 we have that $FG \in H^{-\frac{1}{2},2}(\partial\Sigma)$, as defined above, is well defined. The corresponding estimate follows by the definition and the estimate from Lemma 2.3.6. $\square$

We close the section with the definition of the divergence on submanifolds as well as some important properties of it.

Lemma 2.3.8. Let $\partial\Sigma$ be a $C^{0,1}$-surface and $\{\underline{t}_1, \ldots, \underline{t}_{n-1}\}$ be an orthonormal basis of $T(\partial\Sigma)$. For all $F, G \in H^{\frac{1}{2},2}(\partial\Sigma)$ and $H \in H^{1,\infty}(\partial\Sigma; T(\partial\Sigma))$ we have that $\operatorname{div}_\Sigma(H) \in L^\infty(\partial\Sigma)$ and

$$\sum_{i=1}^n {}_{H^{\frac{1}{2},2}(\partial\Sigma)}\langle F\underline{H}_i, (\nabla_{\partial\Sigma} G)_i\rangle_{H^{-\frac{1}{2},2}(\partial\Sigma)} = -\frac{1}{2}\int_{\partial\Sigma} \operatorname{div}_{\partial\Sigma}(\underline{H}) FG\,dH^{n-1},$$

where the divergence on $\partial\Sigma$ is defined by

$$\operatorname{div}_{\partial\Sigma}(\underline{H}) := \sum_{i=1}^{n-1}\langle \underline{t}_i, \partial_{\underline{t}_i}\underline{H}\rangle.$$

This definition is independent of the basis chosen, see [Alt04, Section 8.2]. Furthermore, if $\partial\Sigma$ is a $C^{1,1}$-surface and $H \in H^{2,\infty}(\partial\Sigma)$ we have $\operatorname{div}_\Sigma(H) \in H^{1,\infty}(\partial\Sigma)$.

Proof. For the proof of the first statement see [GR06, Lemma 2.2.6]. The last statement can be proved analogously, using the fact that we find $\underline{t}_i \in H^{1,\infty}(\partial\Sigma; \mathbb{R}^n)$ for a $C^{1,1}$-surface $\partial\Sigma$, see e.g. [Alt04]. $\square$

2.4 Tools from (Functional) Analysis

The tensor product is the essential tool for implementing stochastic inhomogeneities in Chapter 3. Let H_1 and H_2 be real Hilbert spaces and define for each $\phi_1 \in H_1$ and $\phi_2 \in H_2$ a bilinear form $\phi_1 \otimes \phi_2$ on $H_1 \times H_2$ via

$$\phi_1 \otimes \phi_2(\psi_1, \psi_2) := \langle \phi_1, \psi_1\rangle_{H_1} \cdot \langle \phi_2, \psi_2\rangle_{H_2},$$

for all $\psi_1 \in H_1$ and $\psi_2 \in H_2$. Let Φ be the set consisting of all finite linear combinations of such bilinear forms. We define a scalar product $\langle \cdot, \cdot\rangle_\Phi$ by

$$\langle \phi_1 \otimes \phi_2, \phi_3 \otimes \phi_4\rangle_\Phi := \langle \phi_1, \phi_3\rangle_{H_1} \cdot \langle \phi_2, \phi_4\rangle_{H_2},$$

2.4. TOOLS FROM (FUNCTIONAL) ANALYSIS

for all $\phi_1, \phi_3 \in H_1$, $\phi_2, \phi_4 \in H_2$ and extend it by linearity to Φ. This bilinear form is well defined and positive definite, see [RS72, Proposition II.4.1].

Definition 2.4.1. Let H_1 and H_2 be real Hilbert spaces. We define $H_1 \otimes H_2$ as the completion of Φ under $\langle \cdot, \cdot \rangle_\Phi$. $H_1 \otimes H_2$ is called *tensor product* of H_1 and H_2.

We have the following lemma.

Lemma 2.4.2. If $(\phi_n)_{n \in \mathbb{N}}$ and $(\psi_m)_{m \in \mathbb{N}}$ are orthonormal bases for H_1 and H_2 respectively, then $(\phi_n \otimes \psi_m)_{n,m \in \mathbb{N}}$ is an orthonormal basis of $H_1 \otimes H_2$.

Proof. For the proof of this lemma we refer to [RS72, Proposition II.4.2]. □

Furthermore we have the following useful isomorphisms.

Lemma 2.4.3. Let (M_1, μ_1) and (M_2, μ_2) be measure spaces such that $L^2(M_1, \mu_1)$ and $L^2(M_2, \mu_2)$ are separable and H a separable real Hilbert space. Then:

1. There is an unique isomorphism from $L^2(M_1, \mu_1) \otimes L^2(M_2, \mu_2)$ to $L^2(M_1 \times M_2, \mu_1 \otimes \mu_2)$ such that $f \otimes g \mapsto f \cdot g$, where $f \cdot g(x_1, x_2) = f(x_1) \cdot g(x_2)$ for all $(x_1, x_2) \in M_1 \times M_2$.

2. There is an unique isomorphism from $L^2(M_1, \mu_1) \otimes H$ to $L^2(M_1, \mu_1; H)$ such that $f \otimes \psi \mapsto f \cdot \phi$, where $f \cdot \phi(x) = f(x) \cdot \phi$ for all $x \in M_1$.

3. There is an unique isomorphism from $L^2(M_1 \times M_2, \mu_1 \otimes \mu_2)$ to $L^2(M_1, \mu_1; L^2(M_2, \mu_2))$ such that $(x, y) \mapsto f(x, y)$ is mapped to $x \mapsto f(x, \cdot)$.

Proof. For the proof of this lemma see [RS72, Theorem II.4.10]. □

We go on with the Lax-Milgram Lemma, which is used to provide the solution operator for the inner problem in Chapter 3.

Lemma 2.4.4. Let X be a real Hilbert space and mappings

$$a \ : \ X \times X \to \mathbb{R},$$
$$F \ : \ X \to \mathbb{R},$$

are given such that

$$|a(x,y)| \leq c_{11}\|x\|_X\|y\|_X,$$
$$|a(x,x)| \geq c_{12}\|x\|_X^2,$$
$$|F(x)| \leq c_{13}\|x\|_X,$$

for $0 < c_{11}, c_{12}, c_{13} < \infty$. Then the equation

$$a(x,y) = F(x), \quad \text{for all } x \in X,$$

has one and only one solution $y \in X$. Moreover we have

$$\|y\|_X \leq c_{13}c_{12}.$$

Proof. For the proof see [Alt02, Section 4.1]. □

The following lemma is known as the BLT Theorem and follows from the Hahn-Banach Theorem.

Lemma 2.4.5. Let X, Y be Banach spaces and $T : Z \to Y$ a linear bounded mapping defined on a dense subset $Z \subset X$. Then there exists exactly one linear bounded mapping $\overline{T} : X \to Y$ such that $\overline{T}(z) = T(z)$ for all $z \in Z$. Additionally the operator norm remains invariant, i.e., $\|\overline{T}\|_{L(X;Y)} = \|T\|_{L(Z;Y)}$, where $L(X;Y)$ and $L(Z;Y)$ are the spaces of linear bounded mappings form X to Y and Z to Y, respectively.

Proof. For the proof of this lemma see [RS72, Theorem I.7]. □

We finish this section with two more lemmata which allows us to interchange differentiation with other limiting processes. The first one is about differentiation of sequences.

Lemma 2.4.6. Let $(X, \|\cdot\|_X)$ and $(Y, \|\cdot\|_Y)$ be a Banach spaces such that $Y \subset X$ and $\|f\|_X \leq c_{14}\|f\|_Y$ for all $f \in Y$ where $0 < c_{14} < \infty$. If $(f_n)_{n \in \mathbb{N}} \subset Y$ is a sequence and $g \in X$, $h \in Y$ with $f_n \to g$ in X and $f_n \to h$ in Y then $h = g \in Y$ and $f_n \to g$ in Y.

2.4. TOOLS FROM (FUNCTIONAL) ANALYSIS

Proof. Assume
$$\lim_{n\to\infty} \|f_n - g\|_Y = 0,$$
with $g \in Y$. Because $\|f_n - g\|_Y \leq c_{14}\|f_n - g\|_X$ we have
$$\lim_{n\to\infty} \|f_n - g\|_X = 0,$$
$$\lim_{n\to\infty} \|f_n - h\|_X = 0,$$
for $g, h \in X$. Because X is a Banach space, the limit of a convergent sequence is a unique well defined element of X. Consequently we get $g = h$ in X. □

The following lemma drops as a corollary.

Lemma 2.4.7. Let $\partial\Sigma$ be a $C^{m,\alpha}$-surface, $m \in \mathbb{N}$, $\alpha \in [0,1]$, and $(f_n)_{n\in\mathbb{N}} \subset C^0(\partial\Sigma)$, such that
$$\lim_{n\to\infty} \|f_n - f\|_{C^0(\partial\Sigma)} = 0,$$
for $f \in C^0(\partial\Sigma)$. If $(f_n)_{n\in\mathbb{N}} \subset C^{n,\beta}(\partial\Sigma)$, $n \in \mathbb{N}$, $n \leq m$, $\beta \in [0,1]$, with $n + \beta \leq m + \alpha$ is convergent in $C^{n,\beta}(\partial\Sigma)$, then we have $f \in C^{n,\beta}(\partial\Sigma)$ and
$$\lim_{n\to\infty} \|f_n - f\|_{C^{n,\beta}(\partial\Sigma)} = 0.$$

Proof. This follows as a special case of Lemma 2.4.6 because the spaces are fulfilling the requirements, see Definition 2.2.8. □

The second one deals with differentiation of parameter integrals.

Lemma 2.4.8. Let $f, \frac{\partial f}{\partial x_i} \in C^0\left(B_1^{\mathbb{R}^4}(0)\right)$ and $f, \frac{\partial f}{\partial x_i}$ be equicontinuous, $i \in \{1,2\}$. Furthermore let the parameter integral F be defined by
$$F(x_1, x_2) := \int_{B_1^{\mathbb{R}^2}(0)} f(x_1, x_2, y_1, y_2) d\lambda^2(y_1, y_2),$$
for all $x \in B_1^{\mathbb{R}^2}(0)$. Then we have $F \in C^1\left(B_1^{\mathbb{R}^2}(0)\right)$ with
$$\frac{\partial F}{\partial x_i}(x_1, x_2) = \int_{B_1^{\mathbb{R}^2}(0)} \frac{\partial f}{\partial x_i}(x_1, x_2, y_1, y_2) d\lambda^2(y_1, y_2).$$

Proof. This is a standard result from Analysis for parameter integrals, see e.g. [Heu02]. □

2.5 Some Notions from Probability Theory

In this section we want to give some definitions which are needed in Chapter 3. Only in this section, in Subsection 3.3.3 and in Subsection 3.3.5, Σ denotes a sigma algebra instead of a space domain in $\mathbb{R}^n$.

Definition 2.5.1. Let (Ω, Σ, P) be a probability space and $(\mathbb{R}, \mathbf{B}(\mathbb{R}))$ be the set of real number together with its Borel σ-algebra. Every measurable function

$$X : \Omega \to \mathbb{R},$$

is called *random variable*. For random variables X and Y we define

$$\begin{array}{rll}
\mathrm{E}[X] := & \int_\Omega X\, dP, & \text{if } X \in L^1(\Omega, P), \quad (\textit{expectation value}) \\
\mathrm{cov}(X,Y) := & E[(X - E[Y]) \cdot (Y - E[X])], & \text{if } X, Y \in L^2(\Omega, P), \quad (\textit{covariance}) \\
\mathrm{var}[X] := & \mathrm{cov}(X, X), & \text{if } X \in L^2(\Omega, P), \quad (\textit{variance}) \\
\sigma(X) := & +\sqrt{\mathrm{var}[X]}, & \text{if } X \in L^2(\Omega, P). \quad (\textit{standard deviation})
\end{array}$$

Furthermore we define Gaussian random variables as follows. For details see [Bau02].

Definition 2.5.2. A random variable X having the following distribution

$$P(X \in A) = \frac{1}{\sqrt{2\pi\sigma^2}} \int_A e^{-\frac{(y-\mu)^2}{2\sigma^2}}\, d\lambda^1(y),$$

is called *Gaussian random variable* with expectation value μ and variance $\sigma^2 > 0$. For $\sigma^2 = 0$ we define $X(\omega) := \mu$ for all $\omega \in \Omega$ to be the Gaussian random variable with expectation value μ and variance 0. We set

$$\gamma^{\mu,\sigma^2} := \frac{1}{\sqrt{2\pi\sigma^2}} e^{-\frac{(y-\mu)^2}{2\sigma^2}}\, d\lambda^1(y).$$

One has

$$\int_\Omega H(X)\, dP = \int_\mathbb{R} H(y)\, d\gamma^{\mu,\sigma^2}(y),$$

for all $H \in L^1(\mathbb{R}, \gamma^{\mu,\sigma^2})$ and a Gaussian random variable X with expectation value μ and variance σ^2.

2.6 Function Systems from Geomathematics

In this section we introduce some function systems from geomathematics which will be used in the applications of Chapter 4. Namely this are the system of mass point representations, i.e., the fundamental solutions of the Laplace operator, and the system of inner as well as outer harmonics. We start with a definition.

Definition 2.6.1. Let Σ be an outer $C^{0,1}$-domain. A set of points $(x_k)_{k\in\mathbb{N}} \subset \Sigma$ such that the properties

1. $\mathrm{dist}((x_k)_{k\in\mathbb{N}}, \Sigma) > 0$,

2. Let $F \in C^2(\Sigma)$ with $\Delta F = 0$ in Σ and $F(x_k) = 0$ for all $k \in \mathbb{N}$. Then $F = 0$ in Σ,

are satisfies, is called *fundamental system* in Σ. Furthermore, a set of points $(x_k)_{k\in\mathbb{N}} \subset D$ such that the properties

1. $\mathrm{dist}((x_k)_{k\in\mathbb{N}}, D) > 0$,

2. Let $F \in C^2(D)$ with $\Delta F = 0$ in D and $F(x_k) = 0$ for all $k \in \mathbb{N}$. Then $F = 0$ in D,

are satisfies, is called *fundamental system* in D. For each fundamental system $(x_k)_{k\in\mathbb{N}}$ we define a corresponding system of *mass point representations* given by

$$\left(\frac{1}{|x_k - \cdot|}\right)_{k\in\mathbb{N}}.$$

Let Σ be an outer C^2-domain. Then for example $(x_k)_{k\in\mathbb{N}}$ is a fundamental system if it is a dense subset of each domain Γ such that $\overline{\Gamma}$ is a strict subset of Σ or D, respectively. If $(x_k)_{k\in\mathbb{N}}$ is a dense subset of $\partial\Gamma$ for such a Γ or a dense subset of a parallel surface $\partial\Sigma^{\pm\tau}$, $\tau \in (0, \tau_0]$, then it is also a fundamental system. These examples and definitions are taken from [FM04, Section 2.3.2]. At next we define the systems of inner and outer harmonics.

Definition 2.6.2. The space of *spherical harmonics* of degree n is defined by

$$\mathrm{Harm}_n(\partial B_1^{\mathbb{R}^3}(0)) := \left\{ F\big|_{\partial B_1^{\mathbb{R}^3}(0)} \,\Big|\, F : \mathbb{R}^3 \to \mathbb{R} \text{ is a homogeneous polynomial of degree } n,\, \Delta F = 0 \text{ on } \mathbb{R}^3 \right\}.$$

Let $(Y_{n,k})_{n=0,1,\ldots;k=1,\ldots,2n+1} \subset L^2(\partial B_1^{\mathbb{R}^3}(0))$ such that $\{Y_{n,1},\ldots,Y_{n,2n+1}\}$ forms an orthonormal basis of $\operatorname{harm}_m(\partial B_1^{\mathbb{R}^3}(0))$, for all $n = 0, 1, \ldots$, and $0 < \alpha < \infty$. Then we define the system of *outer harmonics* $\left(H^\alpha_{-n-1,k}\right)_{n=0,1,\ldots;k=1,\ldots,2n+1}$ by

$$H^\alpha_{-n-1,k}(x) := \left(\frac{\alpha}{|x|}\right)^{n+1} \frac{1}{\alpha} Y_{n,k}\left(\frac{x}{|x|}\right),$$

for all $x \in \mathbb{R}^3$. Furthermore, we define the system of *inner harmonics* $\left(H^\alpha_{n,k}\right)_{n=0,1,\ldots;k=1,\ldots,2n+1}$ by

$$H^\alpha_{n,k}(x) := \left(\frac{|x|}{\alpha}\right)^n \frac{1}{\alpha} Y_{n,k}\left(\frac{x}{|x|}\right),$$

for all $x \in \mathbb{R}^3$.

Furthermore we have the following series expansion which will be important in Section 4.7.

Lemma 2.6.3. *Let $0 < \alpha < \infty$. Furthermore let $H^\gamma_{n,j}$ and $H^\gamma_{-n-1,j}$ be given by Definition 2.6.2. Then we have the following series expansion*

$$\frac{1}{|x-y|} = \sum_{n=0}^\infty \frac{4\pi\gamma}{2n+1} \sum_{k=1}^{2n+1} H^\gamma_{n,j}(x) H^\gamma_{-n-1,j}(y),$$

for $x \in B^{\mathbb{R}^3}_\alpha(0)$ and $y \in \mathbb{R}^3 \backslash \overline{B^{\mathbb{R}^3}_\alpha(0)}$.

Proof. This is a standard result about the inner and outer harmonics and is contained in e.g. [FM04]. □

For more details about this function systems and their properties we refer also to [FM04].

Chapter 3

Oblique Boundary Problems

In this chapter we prove existence and uniqueness results for solutions to the outer oblique boundary problem for the Poisson equation under very weak assumptions on boundary, coefficients and inhomogeneities. Main tools are the Kelvin transformation and the solution operator for the regular inner problem, provided in [GR06]. Moreover we prove regularization results for the weak solutions of both, the inner and the outer problem. We investigate the non-admissible direction for the oblique vector field, state results with stochastic inhomogeneities and provide a Ritz-Galerkin approximation. The results are applicable to problems from Geomathematics, see e.g. [Bau04] and [FM02].

3.1 The Inner Regular Oblique Boundary Problem

In this section we present the theory of weak solutions to the regular oblique boundary problem for the Poisson equation for inner domains. Although the weak problem can be formulated for bounded $C^{0,1}$-domains, in order to prove the existence of an unique weak solution we need at least a bounded $C^{1,1}$-domain. Consequently we assume $\Sigma \subset \mathbb{R}^n$ throughout this section to be such a domain, if not stated otherwise. At first we give the definition of the regular oblique boundary problem together with the definition of the classical solution.

Definition 3.1.1. Let Σ be a bounded $C^{1,1}$-domain, $f \in C^0(\Sigma)$, $g, b \in C^0(\partial\Sigma)$ and $\underline{a} \in C^0(\Sigma; \mathbb{R}^n)$ be given, such that

$$|\langle \underline{a}(x), \nu(x) \rangle| > C_1 > 0, \tag{3.1}$$

for all $x \in \partial\Sigma$, where $0 < C_1 < \infty$. Finding a function $u \in C^2(\Sigma) \cap C^1(\overline{\Sigma})$ such that

$$\Delta u = f \quad \text{in} \quad \Sigma, \tag{3.2}$$
$$\langle \underline{a}, \nabla u \rangle + bu = g \quad \text{on} \quad \partial\Sigma, \tag{3.3}$$

is called *inner regular oblique boundary problem* for the *Poisson equation* and u is called *classical solution*.

Because of the condition (3.1) the problem is called regular. It just means that the vector field $\underline{a}$ is non tangential to $\partial\Sigma$ for all $x \in \partial\Sigma$. Now we derive the weak formulation. Due to [Alt02, Lemma 2.21], one has that

$$\Delta u = f \quad \text{in } \Sigma$$

if and only if

$$\int_\Sigma \eta \Delta u \, d\lambda^n = \int_\Sigma \eta f \, d\lambda^n \quad \text{for all } \eta \in C_0^\infty(\Sigma)$$

if and only if

$$\int_\Sigma \eta \Delta u \, d\lambda^n = \int_\Sigma \eta f \, d\lambda^n \quad \text{for all } \eta \in C^\infty(\overline{\Sigma}).$$

Additionally on Σ the following Green formula is valid, see [GT01, Section 2.4]:

$$\int_\Sigma \varphi \Delta \psi \, d\lambda^n + \int_\Sigma \langle \nabla \varphi, \nabla \psi \rangle \, d\lambda^n = \int_{\partial\Sigma} \varphi \frac{\partial \psi}{\partial \nu} \, dH^{n-1},$$

for all $\psi \in C^2(\Sigma) \cap C^1(\overline{\Sigma})$ and $\varphi \in C^\infty(\overline{\Sigma})$. This yields for a classical solution

$$\int_{\partial\Sigma} \eta \frac{\partial u}{\partial \nu} \, dH^{n-1} - \int_\Sigma \langle \nabla \eta, \nabla u \rangle \, d\lambda^n = \int_\Sigma \eta f \, d\lambda^n \quad \text{for all } \eta \in C^\infty(\overline{\Sigma}).$$

Now we transform the boundary condition

$$\langle \underline{a}, \nabla u \rangle + bu = g \quad \text{on} \quad \partial\Sigma,$$

to the form

$$\langle \underline{a}, \nu \rangle \frac{\partial}{\partial \nu} u + \langle \underline{a} - \langle (\underline{a}, \nu) \nu \rangle \cdot \nabla_{\partial\Sigma} u \rangle + bu = g \quad \text{on} \quad \partial\Sigma.$$

Using equation (3.1) we divide by $\langle \underline{a}, \nu \rangle \neq 0$ to get the equivalent boundary condition

$$\frac{\partial}{\partial \nu} u + \left\langle \left(\frac{\underline{a}}{\langle \underline{a}, \nu \rangle} - \nu \right), \nabla_{\partial\Sigma} u \right\rangle + \frac{b}{\langle \underline{a}, \nu \rangle} u = \frac{g}{\langle \underline{a}, \nu \rangle} \quad \text{on} \quad \partial\Sigma.$$

3.1. THE INNER REGULAR OBLIQUE BOUNDARY PROBLEM

Plugging this condition into the equation above, we get the following formulation of the regular oblique boundary problem for the Poisson equation which is equivalent to the formulation given in Definition 3.1.1, see [GR06]. Let the assumptions from Definition 3.1.1 be fulfilled. We want to find a function $u \in C^2(\Sigma) \cap C^1(\overline{\Sigma})$ such that

$$\int_{\partial \Sigma} \eta \left(\frac{g}{\langle \underline{a}, \nu \rangle} - \frac{b}{\langle \underline{a}, \nu \rangle} u - \langle \frac{\underline{a}}{\langle \underline{a}, \nu \rangle} - \nu, \nabla_{\partial \Sigma} u \rangle \right) dH^{n-1} - \int_{\Sigma} \langle \nabla \eta, \nabla u \rangle \, d\lambda^n - \int_{\Sigma} \eta f \, d\lambda^n = 0 \quad \text{for all } \eta \in C^\infty(\overline{\Sigma}).$$

The transformation of the boundary term is shown in Figure 3.

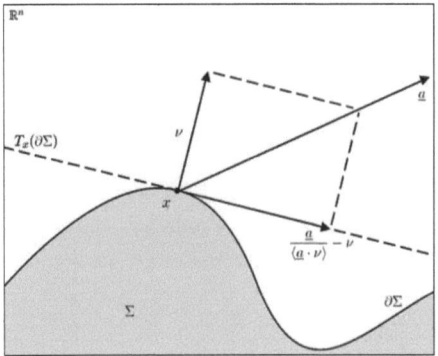

Figure 3: Transformation of the oblique boundary condition

Finally, we are weakening the assumptions on data, coefficients, test function and solution. We give the weak formulation of the inner regular oblique boundary problem to the Poisson equation, summarized in the following definition.

Definition 3.1.2. Let Σ be a bounded $C^{1,1}$-domain, $\underline{a} \in H^{1,\infty}(\partial\Sigma; \mathbb{R}^n)$ fulfilling condition (3.1), $b \in L^\infty(\partial\Sigma)$, $g \in H^{-\frac{1}{2},2}(\partial\Sigma)$ and $f \in (H^{1,2}(\Sigma))'$. We want to find a function $u \in H^{1,2}(\Sigma)$ such that

$$_{H^{\frac{1}{2},2}(\partial\Sigma)} \left\langle \eta, \frac{g}{\langle \underline{a}, \nu \rangle} \right\rangle_{H^{-\frac{1}{2},2}(\partial\Sigma)} - \sum_{i=1}^n {}_{H^{\frac{1}{2},2}(\partial\Sigma)} \left\langle \eta \frac{a_i}{\langle \underline{a}, \nu \rangle} - \nu_i, (\nabla_{\partial\Sigma} u)_i \right\rangle_{H^{-\frac{1}{2},2}(\partial\Sigma)} - \int_{\Sigma} (\nabla \eta \cdot \nabla u) \, d\lambda^n - \int_{\partial\Sigma} \eta \frac{b}{\langle \underline{a}, \nu \rangle} u \, dH^{n-1} - {}_{H^{1,2}(\Sigma)} \langle \eta, f \rangle_{(H^{1,2}(\Sigma))'} = 0, \quad (3.4)$$

for all $\eta \in H^{1,2}(\Sigma)$. Then u is called a *weak solution* of the inner regular oblique boundary problem for the Poisson equation.

All terms in the definition above are well defined, see [GR06]. We have the following existence and uniqueness result for the weak solution.

Theorem 3.1.3. *Let Σ be a bounded $C^{1,1}$-domain, $\underline{a} \in H^{1,\infty}(\partial\Sigma; \mathbb{R}^n)$, fulfilling condition (3.1), and $b \in L^\infty(\partial\Sigma)$ such that:*

$$\operatorname{ess\,inf}_{\partial\Sigma} \left(\frac{b}{\langle \underline{a}, \nu \rangle} - \frac{1}{2} \operatorname{div}_{\partial\Sigma}(\frac{\underline{a}}{\langle \underline{a}, \nu \rangle} - \nu) \right) > 0. \tag{3.5}$$

Then for all $f \in (H^{1,2}(\Sigma))'$ and $g \in H^{-\frac{1}{2},2}(\partial\Sigma)$ there exists one and only one weak solution $u \in H^{1,2}(\Sigma)$ of the inner regular oblique boundary problem for the Poisson equation. Additionally one has for a constant $0 < C_2 < \infty$:

$$\|u\|_{H^{1,2}(\Sigma)} \leq C_2(\|f\|_{(H^{1,2}(\Sigma))'} + \|g\|_{H^{-\frac{1}{2},2}(\partial\Sigma)}). \tag{3.6}$$

Proof. The main tool in this proof is the Lax-Milgram lemma. As mentioned several times before, the proof of this lemma can be found in [GR06]. More detailed proofs can be found in [Ras05]. □

We state some properties of the weak solution in the following remark.

Remark 3.1.4. Condition (3.5) can be transformed into the equivalent form

$$\langle \underline{a}, \nu \rangle b > \frac{1}{2}\big(\langle \underline{a}, \nu \rangle\big)^2 \operatorname{div}_{\partial\Sigma} \left(\frac{\underline{a}}{\langle \underline{a}, \nu \rangle} - \nu \right) \quad H^{n-1}\text{-almost everywhere on } \partial\Sigma.$$

If $\operatorname{div}_{\partial\Sigma} \left(\frac{\underline{a}}{\langle \underline{a}, \nu \rangle} - \nu \right) = 0$ H^{n-1}-almost everywhere on $\partial\Sigma$, one has for H^{n-1}-almost all $x \in \partial\Sigma$ the condition from the classical setting, see for example [GT01] where an existence and uniqueness result for the classical solution is provided. Furthermore for $\underline{a} = \nu$, i.e., the Robin problem, the condition reduces to:

$$b > 0 \quad H^{n-1}\text{-almost everywhere on } \partial\Sigma.$$

Finally we define for each bounded $C^{1,1}$-domain Σ, $\underline{a} \in H^{1,\infty}(\partial\Sigma; \mathbb{R}^n)$ and $b \in L^\infty(\partial\Sigma)$, fulfilling conditions (3.1) and (3.5), a continuous invertible linear solution operator $S_{\underline{a},b}^{\text{in}}$ by

$$S_{\underline{a},b}^{\text{in}} : \big(H^{1,2}(\Sigma)\big)' \times H^{-\frac{1}{2},2}(\partial\Sigma) \to H^{1,2}(\Sigma),$$

3.1. THE INNER REGULAR OBLIQUE BOUNDARY PROBLEM

$$S_{\underline{a},b}^{\text{in}}(f,g) := u,$$

where u is the weak solution provided by Theorem 3.1.3. In addition this means that the inner weak problem is well posed.

We proceed with the following regularization result for the weak solution to the Neumann problem for the Poisson equation.

Lemma 3.1.5. *Let $\Sigma \subset \mathbb{R}^n$ be a bounded C^2-domain. Then for all $f \in L^2(\Sigma)$ and $g \in H^{\frac{1}{2},2}(\partial\Sigma)$ there exists one and only one $u \in H^{2,2}(\Sigma)$ fulfilling*

$$_{H^{\frac{1}{2},2}(\partial\Sigma)}\langle \eta, g\rangle_{H^{-\frac{1}{2},2}(\partial\Sigma)} - \int_\Sigma \langle \nabla\eta, \nabla u\rangle\, d\lambda^n -_{H^{1,2}(\Sigma)}\langle \eta, f\rangle_{(H^{1,2}(\Sigma))'} = 0 \tag{3.7}$$

for all $\eta \in H^{1,2}(\Sigma)$. Moreover the a priori estimate

$$\|u\|_{H^{2,2}(\Sigma)} \leq C_3\Big(\|f\|_{L^2(\Sigma)} + \|g\|_{H^{\frac{1}{2},2}(\partial\Sigma)}\Big), \tag{3.8}$$

holds for a constant $0 < C_3 < \infty$.

Proof. This result is taken from [Dob06, Section 9.4]. □

Using this result we are able to prove an analogous theorem for the regular oblique boundary problem.

Theorem 3.1.6. *Let $\Sigma \subset \mathbb{R}^n$ be a bounded $C^{2,1}$-domain, $\underline{a} \in H^{2,\infty}(\partial\Sigma;\mathbb{R}^n)$ and $b \in H^{1,\infty}(\partial\Sigma)$. Then for all $f \in L^2(\Sigma)$ and $g \in H^{\frac{1}{2},2}(\partial\Sigma)$, the weak solution $u \in H^{1,2}(\Sigma)$ to the inner regular oblique boundary problem for the Poisson equation, provided in Theorem 3.1.3, is even in $H^{2,2}(\Sigma)$. Furthermore we have the a priori estimate*

$$\|u\|_{H^{2,2}(\Sigma)} \leq C_4\Big(\|f\|_{L^2(\Sigma)} + \|g\|_{H^{\frac{1}{2},2}(\partial\Sigma)}\Big). \tag{3.9}$$

for a constant $0 < C_4 < \infty$.

Proof. Assume $f \in L^2(\Sigma)$ and $g \in H^{\frac{1}{2},2}(\partial\Sigma)$. Let $u \in H^{1,2}(\Sigma)$ be the weak solution provided by Theorem 3.1.3. We have

$$0 = {}_{H^{\frac{1}{2},2}(\partial\Sigma)}\left\langle \eta, \frac{g}{\langle\underline{a},\nu\rangle}\right\rangle_{H^{-\frac{1}{2},2}(\partial\Sigma)} - \sum_{i=1}^n {}_{H^{\frac{1}{2},2}(\partial\Sigma)}\left\langle \eta\frac{a_i}{\langle\underline{a},\nu\rangle} - \nu_i, (\nabla_{\partial\Sigma}u)_i\right\rangle_{H^{-\frac{1}{2},2}(\partial\Sigma)}$$

$$\begin{aligned}
&- \int_{\partial\Sigma} \eta \frac{b}{\langle \underline{a},\nu\rangle} u\, dH^{n-1} - \int_{\Sigma} \langle \nabla\eta, \nabla u\rangle\, d\lambda^n -{}_{H^{1,2}(\Sigma)}\langle \eta, f\rangle_{(H^{1,2}(\Sigma))'} \\
={}& {}_{H^{\frac{1}{2},2}(\partial\Sigma)}\Big\langle \eta, \frac{g}{\langle \underline{a},\nu\rangle}\Big\rangle_{H^{-\frac{1}{2},2}(\partial\Sigma)} - \int_{\partial\Sigma} \eta \frac{b}{\langle \underline{a},\nu\rangle} u\, dH^{n-1} \\
&+ \int_{\partial\Sigma} \eta \frac{1}{2}\mathrm{div}_{\partial\Sigma}\!\Big(\frac{\underline{a}}{\langle \underline{a},\nu\rangle} - \nu\Big) u\, dH^{n-1} - \int_{\Sigma} \langle \nabla\eta, \nabla u\rangle\, d\lambda^n - \int_{\Sigma} f\eta\, d\lambda^n \\
={}& \int_{\partial\Sigma} \eta \frac{g}{\langle \underline{a},\nu\rangle} + \eta \frac{b}{\langle \underline{a},\nu\rangle} u + \eta \frac{1}{2}\mathrm{div}_{\partial\Sigma}\!\Big(\frac{\underline{a}}{\langle \underline{a},\nu\rangle} - \nu\Big) u\, dH^{n-1} - \int_{\Sigma} \langle \nabla\eta, \nabla u\rangle\, d\lambda^n \\
&- \int_{\Sigma} f\eta\, d\lambda^n = -\int_{\Sigma} \langle \nabla\eta, \nabla u\rangle\, d\lambda^n - \int_{\Sigma} f\eta\, d\lambda^n \\
&+ \int_{\partial\Sigma} \eta \underbrace{\Big(\frac{g}{\langle \underline{a},\nu\rangle} + \frac{b}{\langle \underline{a},\nu\rangle} u + \frac{1}{2}\mathrm{div}_{\partial\Sigma}\!\Big(\frac{\underline{a}}{\langle \underline{a},\nu\rangle} - \nu\Big) u \Big)}_{=:g^*}\, dH^{n-1},
\end{aligned}$$

for all $\eta \in H^{1,2}(\Sigma)$, using Lemma 2.3.8. Furthermore Lemma 2.3.6 yields that

$$\left\| \frac{g}{\langle \underline{a},\nu\rangle} + \Big(\frac{b}{\langle \underline{a},\nu\rangle} - \frac{1}{2}\mathrm{div}_{\partial\Sigma}\!\Big(\frac{\underline{a}}{\langle \underline{a},\nu\rangle} - \nu\Big) \Big) u \right\|_{H^{\frac{1}{2},2}(\partial\Sigma)} \leq C_5 \Big(\|u\|_{H^{\frac{1}{2},2}(\partial\Sigma)} + \|g\|_{H^{\frac{1}{2},2}(\partial\Sigma)} \Big),$$

for a constant $0 < C_5 < \infty$. Consequently $g^* \in H^{\frac{1}{2},2}(\partial\Sigma)$. So u is the weak solution for the Neumann problem with inhomogeneity $f \in L^2(\Sigma)$ and $g^* \in H^{\frac{1}{2},2}(\partial\Sigma)$. Consequently Lemma 3.1.5 yields $u \in H^{2,2}(\Sigma)$ and the first part of the proof is done. For the estimate we start from inequality (3.8) for the Neumann case above and estimate like follows

$$\begin{aligned}
\|u\|_{H^{2,2}(\Sigma)} &\leq C_4 \Big(\|f\|_{L^2(\Sigma)} + \|g^*\|_{H^{\frac{1}{2},2}(\partial\Sigma)} \Big) \leq C_4 \Big(\|f\|_{L^2(\Sigma)} + C_5 \|g\|_{H^{\frac{1}{2},2}(\partial\Sigma)} \\
&+ C_5 \|u\|_{H^{\frac{1}{2},2}(\partial\Sigma)} \Big) \leq C_3 \Big(\|f\|_{L^2(\Sigma)} + C_5 \|g\|_{H^{\frac{1}{2},2}(\partial\Sigma)} + C_5 c_7 \|u\|_{H^{1,2}(\Sigma)} \Big) \\
&\leq C_4 \Big(\|f\|_{L^2(\Sigma)} + C_5 \|g\|_{H^{\frac{1}{2},2}(\partial\Sigma)} + C_5 c_7 C_2 \Big(\|f\|_{(H^{1,2}(\Sigma))'} + \|g\|_{H^{-\frac{1}{2},2}(\partial\Sigma)} \Big) \Big) \\
&= C_4 \Big(\|f\|_{L^2(\Sigma)} + \|g\|_{H^{\frac{1}{2},2}(\partial\Sigma)} \Big),
\end{aligned}$$

using the trace theorem from Lemma 2.3.3. $\square$

The final lemma of this section verifies that the weak solution is really related to the original problem.

Lemma 3.1.7. Let $u \in H^{2,2}(\Sigma)$ be the weak solution to the inner regular oblique boundary problem for the Poisson equation, provided by Theorem 3.1.6. Then we have

$$\Delta u = f \qquad \lambda^n - \text{almost everywhere in } \Sigma, \tag{3.10}$$

3.2. TRANSFORMATIONS

$$\langle \underline{a}, \nabla u \rangle + bu = g \qquad H^{n-1} - \text{almost everywhere on } \partial \Sigma. \tag{3.11}$$

Such a solution we call *strong solution* to the inner regular oblique boundary problem for the Poisson equation.

Proof. The proof of this lemma can be found in [GR06, Proposition 3.6]. □

3.2 Transformations

In this section we define the transformations which will be needed in order to transform the outer oblique boundary problem for the Poisson equation to a corresponding regular inner problem. Then we will apply the solution operator in order to get a weak solution in the inner domain. This solution will be transformed with help of the Kelvin transformation to a function defined in the outer domain. In the next section we will finally prove that this function solves the outer problem for sufficiently smooth data almost everywhere, which gives the connection to the original problem. The whole procedure is illustrated in the following figure.

Outer problem : $\quad \Sigma \quad \Big| \quad (f,g) \quad \quad \left(\xrightarrow{S_{\underline{a},b}^{\text{out}}} \right) \quad \quad u$

$\quad \quad \quad \quad \quad \downarrow K_\Sigma \quad \quad T_1 \downarrow T_2 \quad \quad \quad \quad \quad \quad \uparrow K$

Inner problem: $\quad \Sigma^K \quad \Big| \quad (T_1(f), T_2(g)) \quad \xrightarrow{S_{T_3(\underline{a}),T_4(b)}^{\text{in}}} \quad v$

We proceed in the following way. First we define the Kelvin transformation K_Σ of the outer domain Σ to a corresponding bounded domain Σ^K. At next the Kelvin transformation K of the solution for the inner problem will be presented. Finally we define the transformations T_1 and T_2 for the inhomogeneities as well as T_3 and T_4 for the coefficients. We will also show that the operators K, T_1 and T_2 are continuous. The consequence is that our solution operator

$$S_{\underline{a},b}^{\text{out}}(f,g) := K \left(S_{T_3(\underline{a}),T_4(b)}^{\text{in}}(T_1(f), T_2(f)) \right),$$

forms a linear and continuous solution operator for the outer problem. Because all main results assume Σ to be at least an outer $C^{1,1}$-domain, we fix Σ in this section as such a domain, if not stated otherwise.

3.2.1 Kelvin Transformation of the Domain

Aim of this subsection is to transform the outer domain Σ to a bounded domain Σ^K. The tool we use is the so called Kelvin transformation K_Σ for domains, which can be found in [Wal71, Section 1.2]. We introduce the Kelvin transformation for outer $C^{1,1}$-domains in the following definition.

Definition 3.2.1. Let Σ be an outer $C^{1,1}$-domain and $x \in \Sigma$ be given. Then we define the *Kelvin transformation* $K_\Sigma(x)$ of x by

$$K_\Sigma(x) := \frac{x}{|x|^2}. \tag{3.12}$$

Furthermore, we define Σ^K as the *Kelvin transformation* of Σ via

$$\Sigma^K := K_\Sigma(\Sigma) \cup \{0\} = \left\{ K_\Sigma(x) \big| x \in \Sigma \right\} \cup \{0\}. \tag{3.13}$$

From this point on, we fix the notation in such a way that Σ^K always means the Kelvin transformation of Σ. The following figure illustrates the Kelvin transformation of Σ.

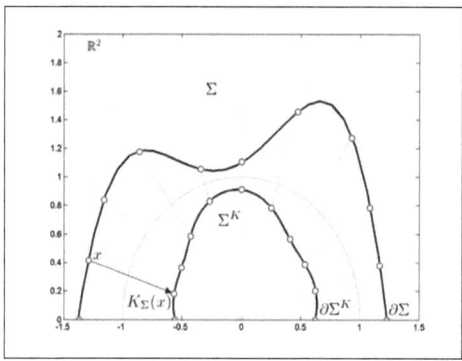

Figure 4: Kelvin transformation of Σ

We have the following lemma about the Kelvin transformation K_Σ.

3.2. TRANSFORMATIONS

Lemma 3.2.2. We have $K_\Sigma \in C^\infty(\mathbb{R}^n \setminus \{0\}; \mathbb{R}^n \setminus \{0\})$ with $K_\Sigma^2 = \mathrm{Id}_{\mathbb{R}^n \setminus \{0\}}$. Furthermore we have

$$|\mathrm{Det}(D(K_\Sigma))(x)| \leq C_6 |x|^{-2n},$$
$$|\partial_i |\mathrm{Det}(D(K_\Sigma))(x)|| \leq C_7 |x|^{-2n-1}$$

for all $x \in \mathbb{R}^n \setminus \{0\}$, $1 \leq i \leq n$, where $0 < C_6, C_7 < \infty$.

Proof. The first two statements about the Kelvin transformations are obvious by the definition. For the first estimate we compute the Jacobian matrix of $K_\Sigma(x) = \frac{x}{|x|^2}$. We have

$$D(K_\Sigma)_{ii} = \frac{1}{|x|^2} - 2\frac{x_i^2}{|x|^4}, \quad i \in \{1, \ldots, n\},$$
$$D(K_\Sigma)_{ij} = -2\frac{x_i x_j}{|x|^4}, \quad i \neq j, i,j \in \{1, \ldots, n\},$$

for all $x \in \mathbb{R}^n \setminus \{0\}$. Now we use the Leibniz formula in order to compute the determinant

$$\begin{aligned}|\mathrm{Det}(D(K_\Sigma))(x)| &= \Big|\sum_{\sigma \in \mathrm{Sym}_n} \mathrm{sign}(\sigma) D(K_\Sigma)_{1\sigma(1)} \cdot \ldots \cdot D(K_\Sigma)_{n\sigma(n)}\Big| \\ &\leq \sum_{\sigma \in \mathrm{Sym}_n} |D(K_\Sigma)_{1\sigma(1)}| \cdot \ldots \cdot |D(K_\Sigma)_{n\sigma(n)}| \\ &\leq \sum_{\sigma \in \mathrm{Sym}_n} \left(\frac{3}{|x|^2}\right)^n = n! 3^n |x|^{-2n},\end{aligned}$$

for all $x \in \mathbb{R}^n \setminus \{0\}$, where Sym_n denotes the set of all permutation of the set $\{1, \ldots, n\}$. This proves the claimed estimate. For the second estimate we obtain that $\mathrm{Det}(D(K_\Sigma))$ is either positive or negative on whole $x \in \mathbb{R}^n \setminus \{0\}$, because K_Σ is an invertible C^1-mapping. Consequently $|\mathrm{Det}(D(K_\Sigma))|$ is differentiable. W.l.o.g. we assume $\mathrm{Det}(D(K_\Sigma))$ to be positive. We have

$$\begin{aligned}\partial_i D(K_\Sigma)_{ii} &= \partial_i \left(\frac{1}{|x|^2} - 2\frac{x_i^2}{|x|^4}\right) = -2\frac{x_i}{|x|^4} + 4\frac{x_i}{|x|^4} + 8\frac{x_i^3}{|x|^6}, \\ \partial_i D(K_\Sigma)_{jj} &= \partial_i \left(\frac{1}{|x|^2} - 2\frac{x_j^2}{|x|^4}\right) = -2\frac{x_i}{|x|^4} + 8\frac{x_i x_j^2}{|x|^6}, \quad j \in \{1, \ldots, n\} \setminus \{i\}, \\ \partial_i D(K_\Sigma)_{ij} &= \partial_i \left(-2\frac{x_i x_j}{|x|^4}\right) = -2\frac{x_j}{|x|^4} + 8\frac{x_i^2 x_j}{|x|^6}, \quad j \in \{1, \ldots, n\} \setminus \{i\}, \\ \partial_i D(K_\Sigma)_{jk} &= \partial_i \left(-2\frac{x_j x_k}{|x|^4}\right) = 8\frac{x_i x_j x_k}{|x|^6}, \quad j \neq k, j,k \in \{1, \ldots, n\} \setminus \{i\},\end{aligned}$$

for all $1 \leq i \leq n$. Finally we compute

$$\begin{aligned}
&|\partial_i|\mathrm{Det}(D(K_\Sigma))(x)|| \\
&= |\partial_i \mathrm{Det}(D(K_\Sigma))(x)| \\
&= \left| \partial_i \sum_{\sigma \in \mathrm{Sym}_n} \mathrm{sign}(\sigma) D(K_\Sigma)_{1\sigma(1)} \cdot \ldots \cdot D(K_\Sigma)_{n\sigma(n)} \right| \\
&\leq \sum_{\sigma \in \mathrm{Sym}_n} \sum_{j=1}^n |\partial_i D(K_\Sigma)_{j\sigma(j)}| \\
&\qquad \cdot |D(K_\Sigma)_{1\sigma(1)}| \cdot \ldots \cdot |D(K_\Sigma)_{j-1\sigma(j-1)}| \cdot |D(K_\Sigma)_{j+1\sigma(j+1)}| \cdot \ldots \cdot |D(K_\Sigma)_{n\sigma(n)}| \\
&\leq n \cdot n! 3^{n-1} |x|^{-2n+2} 14 |x|^{-3} \leq (n+1)! 3^{n+2} |x|^{-2n-1},
\end{aligned}$$

for all $x \in \mathbb{R}^n \setminus \{0\}$, again with help of the Leibniz formula and the product rule of differentiation. $\square$

Furthermore we have the following lemma.

Lemma 3.2.3. Let Σ an outer $C^{2,1}$-domain. Then Σ^K is a bounded $C^{2,1}$-domain. Moreover we have that $\partial \Sigma^K = K_\Sigma(\partial \Sigma)$. Furthermore, if Σ is an outer $C^{1,1}$-domain, we have that Σ^K is a bounded $C^{1,1}$-domain.

Proof. We prove the first statement. Therefore, we have to show that Σ^K is a bounded domain and that there exist an open cover $(U_i^K)_{1 \leq i \leq N}$ of $\partial \Sigma^K$ as well as $C^{2,1}$-mappings $(\Psi_i^K : B_1^{\mathbb{R}^n}(0) \to U_i^K)_{1 \leq i \leq N}$ and $\left(\left(\Psi_i^K\right)^{-1} : U_i^K \to B_1^{\mathbb{R}^n}(0) \right)_{1 \leq i \leq N}$ with $\Psi_i^K(B_1^{\mathbb{R}^{n-1}}(0) \times \{0\}) = U_i^K \cap \partial \Sigma^K$ and $\left(\Psi_i^K\right)^{-1}(U_i^K \cap \partial \Sigma^K) = B_1^{\mathbb{R}^{n-1}}(0) \times \{0\}$, respectively. It is obvious that $\partial \Sigma^K = K_\Sigma(\partial \Sigma)$. Because the Kelvin transformation $K_\Sigma : \mathbb{R}^n \setminus \{0\} \to \mathbb{R}^n \setminus \{0\}$ is continuous and $K_\Sigma^2 = \mathrm{Id}_{\mathbb{R}^n \setminus \{0\}}$ it maps open sets to open sets. So we can define an open cover of $\partial \Sigma^K$ via

$$U_i^K := K_\Sigma(U_i), \quad 1 \leq i \leq N,$$

where $(U_i)_{1 \leq i \leq N}$ is the open cover of $\partial \Sigma$. Furthermore the mappings $\Psi_i^K(x) := K_\Sigma(\Psi_i(x))$ and $\left(\Psi_i^K\right)^{-1}(x) := \Psi_i^{-1}(K_\Sigma(x))$ fulfill $\Psi_i^K \circ \left(\Psi_i^K\right)^{-1} = \mathrm{Id}_{U_i}$ and $\left(\Psi^K\right)_i^{-1} \circ \Psi_i^K = \mathrm{Id}_{B_1^{\mathbb{R}^n}(0)}$. We have $\Psi_i(B_1^{\mathbb{R}^{n-1}}(0) \times \{0\}) = U_i^K \cap \partial \Sigma^K$ and $\Psi_i^{-1}(U_i^K \cap \partial \Sigma^K) = B_1^{\mathbb{R}^{n-1}}(0) \times \{0\}$, respectively. It is left to verify is that these are $C^{2,1}$-mappings. We use the isomorphisms

$$C^{2,1}(B_1^{\mathbb{R}^n}(0); U_i) \cong H^{3,\infty}(B_1^{\mathbb{R}^n}(0); U_i),$$

3.2. TRANSFORMATIONS

$$C^{2,1}(U_i; B_1^{\mathbb{R}^n}(0)) \cong H^{3,\infty}(U_i; B_1^{\mathbb{R}^n}(0)),$$
$$C^{2,1}(B_1^{\mathbb{R}^n}(0); U_i^K) \cong H^{3,\infty}(B_1^{\mathbb{R}^n}(0); U_i^K),$$
$$C^{2,1}(U_i^K; B_1^{\mathbb{R}^n}(0)) \cong H^{3,\infty}(U_i^K; B_1^{\mathbb{R}^n}(0)).$$

for $i \in \{1, \ldots, N\}$, see Lemma 2.2.5. Additionally we have

$$\text{dist}(\partial \Sigma, 0) > 0,$$
$$\text{dist}(\partial \Sigma^K, 0) > 0.$$

This yields that $KT_D : \Sigma^K \to \Sigma$ is a bounded C^∞-diffeomorphism. Now we can apply chain and product rule for functions from Sobolev spaces, see Lemma 2.2.4, to get

$$(\Psi_i^K)_{1 \le i \le N} \subset H^{3,\infty}(B_1^{\mathbb{R}^n}(0); U_i^K) \cong C^{2,1}(B_1^{\mathbb{R}^n}(0); U_i^K),$$
$$\left((\Psi_i^K)^{-1}\right)_{1 \le i \le N} \subset H^{3,\infty}(U_i^K; B_1^{\mathbb{R}^n}(0)) \cong C^{2,1}(U_i^K; B_1^{\mathbb{R}^n}(0)),$$

for $i \in \{1, \ldots, N\}$. In the same way, the statement for outer $C^{1,1}$-domains Σ can be proved, and the proof is done. □

Remark 3.2.4. There are geometric situations in which $\partial \Sigma^K$ can be computed easily. For example if $\partial \Sigma$ is a sphere around the origin with radius R, then $\partial \Sigma^K$ is a sphere around the origin with radius R^{-1}. Furthermore, if $\partial \Sigma \subset \mathbb{R}^2$ is an ellipse with semi axes a and b around the origin, then $\partial \Sigma^K$ is also an ellipse around the origin with semi axes a^{-1} and b^{-1}.

3.2.2 Kelvin Transformation of the Solution

In this subsection we introduce the operator K. This is the so called Kelvin transformation for functions. It transforms a given function u, defined in Σ^K, to a function $K(u)$, defined in Σ. In addition, it preserves some properties of the original function. We will state some of these properties. So, after this subsection it will be clear why we choose exactly this transformation. It will also be clear how we have to choose the transformations $T_1, \ldots, T_4$ in Subsection 3.2.3. We start with a definition.

Definition 3.2.5. Let Σ be an outer $C^{1,1}$-domain and u be a function defined on Σ^K. Then we define the *Kelvin transformation* $K(u)$ of u, which is a function defined on Σ, via

$$K(u)(x) := \frac{1}{|x|^{n-2}} u\left(\frac{x}{|x|^2}\right), \tag{3.14}$$

for all $x \in \Sigma$.

Important is that this transformation acts as a multiplier when applying the Laplace operator. Note that $-(n-2)$ is the only exponent for $|x|$ which has this property. We have the following lemma.

Lemma 3.2.6. Let Σ be an outer $C^{1,1}$-domain and $u \in C^2(\Sigma^K)$. Then we have $K(u) \in C^2(\Sigma)$ with
$$\Delta(K(u))(x) = \frac{1}{|x|^{n+2}}(\Delta u)(\frac{x}{|x|^2}),$$
for all $x \in \Sigma$.

Proof. For the proof of this lemma see [Wal71, Paragraph 2]. □

As already mentioned above we will apply K to functions from $H^{1,2}(\Sigma^K)$. So we want to find a normed function space $(V, \|\cdot\|_V)$ such that
$$K : H^{1,2}(\Sigma^K) \to V$$
defines a continuous operator. It turns out that the weighted Sobolev space $H^{1,2}_{\frac{1}{|x|^2},\frac{1}{|x|}}(\Sigma)$ is a suitable choice. We have the following result.

Lemma 3.2.7. Let Σ be an outer $C^{1,1}$-domain. For $u \in H^{1,2}(\Sigma^K)$ let $K(u)$ be defined by equation (3.14) for all $x \in \Sigma$. Then we have that
$$K : H^{1,2}(\Sigma^K) \to H^{1,2}_{\frac{1}{|x|^2},\frac{1}{|x|}}(\Sigma)$$
is a continuous Operator, i.e., K is linear with
$$\|K(u)\|_{H^{1,2}_{\frac{1}{|x|^2},\frac{1}{|x|}}(\Sigma)} \leq C_8 \|u\|_{H^{1,2}(\Sigma^K)}, \tag{3.15}$$
for all $u \in H^{1,2}(\Sigma^K)$, where $0 < C_8 < \infty$. Moreover K is injective.

Proof. By the definition of K injectivity is obvious. Now let $u \in H^{1,2}(\Sigma^K)$ be given. We have to prove
$$\|K(u)\|_{H^{1,2}_{\frac{1}{|x|^2},\frac{1}{|x|}}(\Sigma)} \leq C_8 \|u\|_{H^{1,2}(\Sigma^K)},$$

3.2. TRANSFORMATIONS

where the constant $0 < C_8 < \infty$ is independent of u. With help of the chain and product rule of differentiation for Sobolev spaces, see Lemma 2.2.4, $K(u)$ is a weakly differentiable function. Due to the transformation formula for integrals we have

$$\|K(u)\|^2_{H^{1,2}_{\frac{1}{|x|^2},\frac{1}{|x|}}(\Sigma)}$$

$$= \|K(u)\|^2_{L^2_{\frac{1}{|x|^2}}(\Sigma)} + \sum_{i=1}^{n} \|\partial_i(K(u))\|^2_{L^2_{\frac{1}{|x|}}(\Sigma)}$$

$$= \int_\Sigma \left(\frac{1}{|x|^{n-2}}u(\frac{x}{|x|^2})\right)^2 \frac{1}{|x|^4}d\lambda^n(x) + \sum_{i=1}^{n} \int_\Sigma \left(\partial_i\left(\frac{1}{|x|^{n-2}}u(\frac{x}{|x|^2})\right)\right)^2 \frac{1}{|x|^2}d\lambda^n(x)$$

$$= \int_\Sigma \frac{1}{|x|^{2n}}u^2(\frac{x}{|x|^2})d\lambda^n(x) + \sum_{i=1}^{n}\int_\Sigma \left((2-n)^2 \frac{x_i^2}{|x|^{2n}}u^2(\frac{x}{|x|^2}) + (\partial_i u)^2(\frac{x}{|x|^2})\frac{1}{|x|^{2n}}\right.$$

$$+ 4\sum_{j=1}^{n}(\partial_j u)^2(\frac{x}{|x|^2})\frac{x_i^2 x_j^2}{|x|^{2n+4}} + 2(2-n)\frac{x_i}{|x|^{2n}}u(\frac{x}{|x|^2})(\partial_i u)(\frac{x}{|x|^2})$$

$$-4(2-n)\sum_{j=1}^{n}(\partial_j u)(\frac{x}{|x|^2})u(\frac{x}{|x|^2})\frac{x_i^2 x_j}{|x|^{2n+2}} - 4\sum_{j=1}^{n}(\partial_j u)(\frac{x}{|x|^2})(\partial_i u)(\frac{x}{|x|^2})\frac{x_i x_j}{|x|^{2n+2}}$$

$$\left. + 4\sum_{j=1}^{n}\sum_{m=1, m\neq j}^{n}(\partial_j u)(\frac{x}{|x|^2})(\partial_m u)(\frac{x}{|x|^2})\frac{x_i^2 x_j x_m}{|x|^{2n+4}}\right)\frac{1}{|x|^2}d\lambda^n(x)$$

$$\leq \int_\Sigma \frac{1}{|x|^{2n}}u^2(\frac{x}{|x|^2})d\lambda^n(x)$$

$$+ \sum_{i=1}^{n}\int_\Sigma \left((2-n)^2\frac{1}{|x|^{2n}}u^2(\frac{x}{|x|^2}) + (\partial_i u)^2(\frac{x}{|x|^2})\frac{1}{|x|^{2n+2}}\right.$$

$$+ 4\sum_{j=1}^{n}(\partial_j u)^2(\frac{x}{|x|^2})\frac{1}{|x|^{2n+2}} + 2(2-n)\frac{1}{|x|^{2n+1}}u(\frac{x}{|x|^2})(\partial_i u)(\frac{x}{|x|^2})$$

$$-4(2-n)\sum_{j=1}^{n}(\partial_j u)(\frac{x}{|x|^2})u(\frac{x}{|x|^2})\frac{1}{|x|^{2n+1}} - 4\sum_{j=1}^{n}(\partial_j u)(\frac{x}{|x|^2})(\partial_i u)(\frac{x}{|x|^2})\frac{1}{|x|^{2n+2}}$$

$$\left. + 4\sum_{j=1}^{n}\sum_{m=1,m\neq j}^{n}(\partial_j u)(\frac{x}{|x|^2})(\partial_m u)(\frac{x}{|x|^2})\frac{1}{|x|^{2n+2}}\right)d\lambda^n(x)$$

$$= \int_{\Sigma^K}|y|^{2n}u^2(y)|\mathrm{Det}(D(K_\Sigma(y)))|d\lambda^n(y) +$$

$$\sum_{i=1}^{n}\int_{\Sigma^K}\left((2-n)^2|y|^{2n}u^2(y) + (\partial_i u)^2(y)|y|^{2n}\right.$$

$$\begin{aligned}&+4\sum_{j=1}^{n}(\partial_{j}u)^{2}(y)|y|^{2n+2}+2(2-n)|y|^{2n+1}u(y)(\partial_{i}u)(y)\\&-4(2-n)\sum_{j=1}^{n}(\partial_{j}u)(y)u(y)|y|^{2n+1}-4\sum_{j=1}^{n}(\partial_{j}u)(y)(\partial_{i}u)(y)|y|^{2n+2}\\&+4\sum_{j=1}^{n}\sum_{m=1,m\neq j}^{n}(\partial_{j}u)(y)(\partial_{m}u)(y)|y|^{2n+2}\bigg)\mathrm{Det}(D(K_{\Sigma}(y)))|d\lambda^{n}(y)\\\leq\ &\|u\|_{L^{2}(\Sigma^{K})}^{2}+\\&\sum_{i=1}^{n}\int_{\Sigma^{K}}\bigg((2-n)^{2}u^{2}(y)+(\partial_{i}u)^{2}(y)|y|^{2}\\&+4\sum_{j=1}^{n}(\partial_{j}u)^{2}(y)|y|^{2}+2(2-n)|y||u(y)||(\partial_{i}u)(y)|\\&-4(2-n)\sum_{j=1}^{n}|(\partial_{j}u)(y)||u(y)||y|-4\sum_{j=1}^{n}|(\partial_{j}u)(y)||(\partial_{i}u)(y)||y|^{2}\\&+4\sum_{j=1}^{n}\sum_{m=1,m\neq j}^{n}|(\partial_{j}u)(y)||(\partial_{m}u)(y)||y|^{2}\bigg)d\lambda^{n}(y)\\\leq\ &C_{8}\|u\|_{H^{1,2}(\Sigma^{K})}^{2}\end{aligned}$$

where C_8 only depends on n and $\sup_{y\in\Sigma^K}\{|y|\}<\infty$. $\square$

3.2.3 Transformation of Inhomogeneities and Coefficients

This subsection provides the remaining transformations $T_1,\ldots,T_4$. In the first part we treat T_1, which transforms the inhomogeneity f of the outer problem in Σ to an inhomogeneity of the corresponding inner problem in Σ^K. Assume f to be a function defined on Σ. We want to define the function $T_1(f)$ on Σ^K, such that

$$\Delta u(x)=T_1(f)(x),\quad x\in\Sigma^K, \qquad (3.16)$$

implies that

$$\Delta(K(u))(y)=f(y),\quad y\in\Sigma. \qquad (3.17)$$

Using Lemma 3.2.6 we are able to define T_1 for functions defined on Σ as follows.

3.2. TRANSFORMATIONS

Definition 3.2.8. Let Σ be an outer $C^{1,1}$-domain and f be a function defined in Σ. Then we define a function $T_1(f)$ on Σ^K by

$$T_1(f)(x) := \frac{1}{|x|^{n+2}} f(\frac{x}{|x|^2}), \tag{3.18}$$

for all $x \in \Sigma^K \setminus \{0\}$ and $T_1(f)(0) = 0$.

The next lemma summarizes the most important properties of this transformation.

Lemma 3.2.9. T_1 is well defined and fulfills the relation described by equations (3.16) and (3.17). Furthermore, T_1 defines a linear continuous isomorphism

$$T_1 : L^2_{|x|^2}(\Sigma) \to L^2(\Sigma^K),$$

with $(T_1)^{-1} = T_1$.

Proof. The first statement follows immediately by using Lemma 3.2.6. In order to prove that T_1 is well defined we will show

$$\|T_1(f)\|_{L^2(\Sigma^K)} \leq C_6 \|f\|_{L^2_{|x|^2}(\Sigma)},$$

for all $f \in L^2_{|x|^2}(\Sigma)$ and

$$\|T_1(f)\|_{L^2_{|x|^2}(\Sigma)} \leq C_6 \|f\|_{L^2(\Sigma^K)},$$

for all $f \in L^2_{|x|^2}(\Sigma)$. C_6 is the constant from Lemma 3.2.2. We have

$$\|T_1(f)\|^2_{L^2(\Sigma^K)} = \int_{\Sigma^K} \left(\frac{1}{|x|^{n+2}} f(\frac{x}{|x|^2})\right)^2 d\lambda^n(x) = \int_{\Sigma^K} \frac{1}{|x|^{2n+4}} f^2(\frac{x}{|x|^2}) d\lambda^n(x)$$

$$= \int_\Sigma |y|^{2n+4} f^2(y) |\text{Det}(D(K_\Sigma))(y)| d\lambda^n(y) \leq C_6 \int_\Sigma |y|^{2n+4} f^2(y) |y|^{-2n} d\lambda^n(y)$$

$$= C_6 \int_{\Sigma^K} f^2(y) |y|^4 d\lambda^n(y) = C_6 \|f\|^2_{L^2_{|x|^2}(\Sigma)}.$$

The other direction can be proved in a completely analogous way. Bijectivity is obvious by the definition, consequently T_1 defines a continuous isomorphism and the proof is complete. $\square$

We want to generalize our inhomogeneities in a way similar to the inner problem. This means we have to identify a normed vector space $(W, \|\cdot\|_W)$, such that

$$T_1 : W \to \left(H^{1,2}(\Sigma^K)\right)',$$

defines a linear continuous operator. Additionally, because of Lemma 3.2.9, we want to end up with a Gelfand triple
$$U \subset L^2_{|x|^2}(\Sigma) \subset W.$$
Consequently $L^2_{|x|^2}(\Sigma)$ should be a dense subspace. Using the following lemma, we will be able to prove that the space $\left(H^{1,2}_{|x|^2,|x|^3}(\Sigma)\right)'$ is a suitable choice.

Lemma 3.2.10. Let Σ be an outer $C^{1,1}$-domain. The operator $J: H^{1,2}(\Sigma^K) \to H^{1,2}_{|x|^2,|x|^3}(\Sigma)$ defined by
$$J(u)(x) := u(\frac{x}{|x|^2}) \cdot |x|^{n-2} \cdot |\mathrm{Det}(D(K_\Sigma))(x)|, \quad x \in \Sigma,$$
for $u \in H^{1,2}(\Sigma^K)$, is a continuous linear operator.

Proof. First of all, we want to mention that J is well defined. $|\mathrm{Det}(D(K_\Sigma))| > 0$ in $\overline{\Sigma}$ and consequently it is differentiable, see Lemma 3.2.2. Additionally, the product rule as well as the chain rule of differentiation is available for weakly differentiable functions, see Lemma 2.2.4. So $J(u)$ defines a weakly differentiable function in Σ. The operator J is obviously linear. So it is left to prove
$$\|J(u)\|_{H^{1,2}_{|x|^2,|x|^3}(\Sigma)} \leq C_9 \|u\|_{H^{1,2}(\Sigma^K)},$$
for a constant $0 < C_9 < \infty$. The estimate
$$\|J(u)\|^2_{L^2_{|x|^2}(\Sigma)} \leq C_6 \|u\|^2_{H^{1,2}(\Sigma^K)},$$
can be derived analogous to the estimate in the proof of the Lemma 3.2.9. So, it remains to estimate the weak derivatives for $1 \leq i \leq n$. We do this in the following way
$$\|\partial_i(J(u))\|^2_{L^2_{|x|^3}(\Sigma)}$$
$$= \int_\Sigma \left(\partial_i \left(u(\frac{x}{|x|^2}) \cdot |x|^{n-2} \cdot |\mathrm{Det}(D(K_\Sigma))(x)| \right) \right)^2 |x|^6 d\lambda^n(x)$$
$$= \int_\Sigma \left(\partial_i \left(u(\frac{x}{|x|^2}) \cdot |x|^{n-2} \right) \cdot |\mathrm{Det}(D(K_\Sigma))(x)| \right.$$
$$\left. + u(\frac{x}{|x|^2}) \cdot |x|^{n-2} \cdot \partial_i |\mathrm{Det}(D(K_\Sigma))(x)| \right)^2 |x|^6 d\lambda^n(x)$$
$$= \int_\Sigma \left((n-2)|x|^{n-4} x_i u(\frac{x}{|x|^2}) \cdot |\mathrm{Det}(D(K_\Sigma))(x)| \right.$$

3.2. TRANSFORMATIONS

$$+|x|^{n-4}\partial_i u(\frac{x}{|x|^2}) \cdot |\text{Det}(D(K_\Sigma))(x)|$$

$$-2\sum_{j=1}^{n}\partial_j u(\frac{x}{|x|^2})x_i x_j |x|^{n-6} \cdot |\text{Det}(D(K_\Sigma))(x)|$$

$$+u(\frac{x}{|x|^2}) \cdot |x|^{n-2} \cdot \partial_i|\text{Det}(D(K_\Sigma))(x)|\Big)^2 |x|^6 d\lambda^n(x)$$

$$\leq (C_6 + C_7)^2 \int_\Sigma \Big(|n-2||x|^{-n-3}|u|(\frac{x}{|x|^2}) + |x|^{-n-4}|\partial_i u|(\frac{x}{|x|^2})$$

$$+2\sum_{j=1}^{n}|x|^{-n-4}|\partial_j u|(\frac{x}{|x|^2}) + |x|^{-n-3}|u|(\frac{x}{|x|^2})\Big)^2 |x|^6 d\lambda^n(x)$$

$$\leq (C_6 + C_7)^2 \int_\Sigma \Big(n|x|^{-n}|u|(\frac{x}{|x|^2}) + 3\sum_{j=1}^{n}|x|^{-n-1}|\partial_j u|(\frac{x}{|x|^2})\Big)^2 d\lambda^n(x)$$

$$\leq (C_6 + C_7)^2 \Big(\int_\Sigma n^2 |x|^{-2n}|u|^2(\frac{x}{|x|^2})d\lambda^n(x) + 3\sum_{j=1}^{n}\int_\Sigma |x|^{-2n-2}|\partial_j u|^2(\frac{x}{|x|^2})d\lambda^n(x)\Big)^2$$

$$= (C_6 + C_7)^2 \Big(\int_{\Sigma^K} n^2 |y|^{2n}|u|^2(y)|\text{Det}(D(K_\Sigma))(y)|d\lambda^n(y)$$

$$+3\sum_{j=1}^{n}\int_\Sigma |y|^{2n+2}|\partial_j u|^2(y)|\text{Det}(D(K_\Sigma))(y)|d\lambda^n(y)\Big)^2$$

$$\leq (C_6 + C_7)^2 (n^4 + 9n^2 \sup\{|y|^2 | y \in \Sigma^K\}) \|u\|^2_{H^{1,2}(\Sigma^K)}.$$

□

Now we are able to extend our definition of T_1 to elements of $\left(H^{1,2}_{|x|^2,|x|^3}(\Sigma)\right)'$. Before we go to the next lemma, it might be useful to recall the Gelfand triple, given in Remark 2.2.7 (iii) by

$$H^{1,2}_{|x|^2,|x|^3}(\Sigma) \subset L^2_{|x|^2}(\Sigma) \subset \left(H^{1,2}_{|x|^2,|x|^3}(\Sigma)\right)'.$$

Lemma 3.2.11. Let Σ be an outer $C^{1,1}$-domain. We define a continuous linear operator

$$T_1 : L^2_{|x|^2}(\Sigma) \to \left(H^{1,2}(\Sigma^K)\right)',$$

by

$$(T_1(f))(h) := \int_{\Sigma^K} (T_1(f))(y) h(y) d\lambda^n(y), \quad h \in H^{1,2}(\Sigma^K), \tag{3.19}$$

for $f \in L^2_{|x|^2}(\Sigma)$, where $L^2_{|x|^2}(\Sigma)$ is equipped with the norm $\|\cdot\|_{\left(H^{1,2}_{|x|^2,|x|^3}(\Sigma)\right)'}$. There exists a linear bounded continuation

$$T_1 : \left(H^{1,2}_{|x|^2,|x|^3}(\Sigma)\right)' \to \left(H^{1,2}(\Sigma^K)\right)',$$

i.e., we are able to find $0 < C_{10} < \infty$ such that

$$\|T_1(f)\|_{(H^{1,2}(\Sigma^K))'} \leq C_{10} \|f\|_{\left(H^{1,2}_{|x|^2,|x|^3}(\Sigma)\right)'}, \tag{3.20}$$

for all $f \in \left(H^{1,2}_{|x|^2,|x|^3}(\Sigma)\right)'$.

Proof. Obviously the operator T_1, as defined above, is well defined on the set $L^2_{|x|^2}(\Sigma) \subset \left(H^{1,2}_{|x|^2,|x|^3}(\Sigma)\right)'$ because $T_1 : L^2_{|x|^2}(\Sigma) \to L^2(\Sigma^K)$, see Lemma 3.2.9, and

$$\|T_1(f)\|_{(H^{1,2}(\Sigma^K))'} \leq \|T_1(f)\|_{L^2(\Sigma^K)},$$

for $f \in L^2_{|x|^2}(\Sigma)$. Moreover it is linear on this dense subset. We will show that

$$\|T_1(f)\|_{(H^{1,2}(\Sigma^K))'} \leq C_9 \|f\|_{\left(H^{1,2}_{|x|^2,|x|^3}(\Sigma)\right)'},$$

for all $f \in L^2_{|x|^2}(\Sigma)$. Then the BLT theorem, see Lemma 2.4.5, gives us an unique linear continuation to $\left(H^{1,2}_{|x|^2,|x|^3}(\Sigma)\right)'$ with same bound. We compute for $f \in L^2_{|x|^2}(\Sigma)$ and $h \in H^{1,2}(\Sigma^K)$

$$|(T_1(f))(h)| = \left|\int_{\Sigma^K} (T_1(f))(x) h(x) d\lambda^n(x)\right| = \left|\int_{\Sigma^K} \frac{1}{|x|^{n+2}} f\left(\frac{x}{|x|^2}\right) h(x) d\lambda^n(x)\right|$$

$$= \left|\int_\Sigma |y|^{n+2} f(y) h\left(\frac{y}{|y|^2}\right) |\text{Det}(D(K_\Sigma))(y)| d\lambda^n(y)\right| = \left|\int_\Sigma f(y) (J(h))(y) |x|^4 d\lambda^n(y)\right|$$

$$\leq \|f\|_{\left(H^{1,2}_{|x|^2,|x|^3}(\Sigma)\right)'} \|(J(h))(y)\|_{H^{1,2}_{|x|^2,|x|^3}(\Sigma)} \leq C_9 \|f\|_{\left(H^{1,2}_{|x|^2,|x|^3}(\Sigma)\right)'} \|h\|_{H^{1,2}(\Sigma^K)},$$

using Lemma 3.2.10. But this yields

$$\|T_1(f)\|_{(H^{1,2}(\Sigma^K))'} = \sup_{h \in H^{1,2}(\Sigma^K), \|h\|_{H^{1,2}(\Sigma^K)}=1} |(T_1(f))(h)| \leq C_9 \|f\|_{\left(H^{1,2}_{|x|^2,|x|^3}(\Sigma)\right)'}.$$

$\square$

The second part of this subsection contains the transformations for the boundary inhomogeneity g and the coefficients $\underline{a}$ and b. This means we want to find transformations T_2, T_3 and T_4 such that

$$\langle (T_3(\underline{a}))(x), \nabla u(x)\rangle + (T_4(b))(x) u(x) = (T_2(g))(x), \tag{3.21}$$

3.2. TRANSFORMATIONS

for all $x \in \partial \Sigma^K$, yields that

$$\langle \underline{a}(y), \nabla((K(u))(y)) \rangle + b(y)u(x) = g(y), \tag{3.22}$$

for all $y \in \partial \Sigma^K$. We start with the transformation $T_2(g)$ of g.

Definition 3.2.12. Let Σ be an outer $C^{1,1}$-domain and g be a function defined on $\partial \Sigma$. Then we define a function $T_2(g)$ on $\partial \Sigma^K$ by

$$(T_2(g))(x) := g(\frac{x}{|x|^2}), \quad x \in \partial \Sigma^K. \tag{3.23}$$

Again we recall a Gelfand triple from Remark 2.2.7 (iii), namely

$$H^{\frac{1}{2},2}(\partial \Sigma) \subset L^2(\partial \Sigma) \subset H^{-\frac{1}{2},2}(\partial \Sigma).$$

T_2 has the following properties.

Lemma 3.2.13. Let Σ be an outer $C^{1,1}$-domain. We have that

$$T_2 : L^2(\partial \Sigma) \to L^2(\partial \Sigma^K),$$
$$T_2 : H^{\frac{1}{2},2}(\partial \Sigma) \to H^{\frac{1}{2},2}(\partial \Sigma^K),$$

define linear, bounded isometries with $(T_2)^{-1} = T_2$. Moreover we define a continuous linear operator

$$T_2 : L^2(\partial \Sigma) \to H^{-\frac{1}{2},2}(\partial \Sigma^K),$$

by

$$(T_2(g))(h) := \int_{\partial \Sigma^K} T_2(g)(y) h(y) \, dH^{n-1}(y), \quad h \in H^{-\frac{1}{2},2}(\partial \Sigma) \tag{3.24}$$

for $g \in L^2(\partial \Sigma)$, where $L^2(\partial \Sigma)$ is equipped with the norm $\|\cdot\|_{H^{-\frac{1}{2},2}(\partial \Sigma)}$. Hence there exists a continuous continuation

$$T_2 : H^{-\frac{1}{2},2}(\partial \Sigma) \to H^{-\frac{1}{2},2}(\partial \Sigma^K),$$

i.e., we have for a constant $0 < C_{11} < \infty$

$$\|T_2(g)\|_{H^{-\frac{1}{2},2}(\partial \Sigma^K)} \leq C_{11} \|g\|_{H^{-\frac{1}{2},2}(\partial \Sigma)}, \tag{3.25}$$

for all $g \in H^{-\frac{1}{2},2}(\partial \Sigma)$.

Proof. Obviously $(T_2)^{-1} = T_2$ and the operators are linear. At first we consider T_2 applied to $g \in L^2(\partial\Sigma)$. Let $(\Psi_i)_{1\leq i\leq N}$ be the mappings from $\partial\Sigma$ and $\left(\frac{\Psi_i}{|\Psi_i|^2}\right)_{1\leq i\leq N}$ be those from $\partial\Sigma^K$. Furthermore let $(w_i)_{1\leq i\leq N}$ be a C^∞-partition of unity on $\partial\Sigma$ corresponding to $(U_i)_{1\leq i\leq N}$. Then $\left(w_i^K\right)_{1\leq i\leq N}$ defined by

$$w_i^K(x) := w_i(\frac{x}{|x|^2}), \quad x \in U_i^K,$$

for $1 \leq i \leq N$, is a C^∞-partition of unity on $\partial\Sigma^K$, corresponding to the open cover $\left(U_i^K\right)_{1\leq i\leq N}$. Details can be found in the proof of Lemma 3.2.3. Now we compute

$$\|T_2(g)\|^2_{L^2(\partial\Sigma^K)} = \sum_{i=1}^N \|w_i(\Psi_i) \cdot (T_2(g))(\Psi_i)\|^2_{L^2(\mathbb{R}^{n-1})}$$

$$= \sum_{i=1}^N \|w_i^K(\frac{\Psi_i}{|\Psi_i|^2}) \cdot g(\frac{\Psi_i}{|\Psi_i|^2})\|^2_{L^2(\mathbb{R}^{n-1})} = \sum_{i=1}^N \|w_i^K(\Psi_i^K) \cdot g(\Psi_i^K)\|^2_{L^2(\mathbb{R}^{n-1})} = \|g\|^2_{L^2(\partial\Sigma)}.$$

For T_2 applied to $H^{\frac{1}{2},2}(\partial\Sigma)$ we compute the norm in the following way, using an equivalent definition of the $H^{\frac{1}{2},2}(\partial\Sigma^K)$ and $H^{\frac{1}{2},2}(\partial\Sigma)$ norms. [Dob06, Section 6.10 and Definition 9.39] yields

$$\|g\|^2_{H^{\frac{1}{2},2}(\partial\Sigma^K)} = \|g\|^2_{L^2(\partial\Sigma^K)}$$
$$+ \sum_{i=1}^N \int_{B_1^{\mathbb{R}^{n-1}}(0)} \int_{B_1^{\mathbb{R}^{n-1}}(0)} \frac{|g(\Psi_i^K(x,0)) - g(\Psi_i^K(y,0))|^2}{|x-y|^n} d\lambda^{n-1}(x) d\lambda^{n-1}(y),$$

$$\|g\|^2_{H^{\frac{1}{2},2}(\partial\Sigma)} = \|g\|^2_{L^2(\partial\Sigma)}$$
$$+ \sum_{i=1}^N \int_{B_1^{\mathbb{R}^{n-1}}(0)} \int_{B_1^{\mathbb{R}^{n-1}}(0)} \frac{|g(\Psi_i(x,0)) - g(\Psi_i(y,0))|^2}{|x-y|^n} d\lambda^{n-1}(x) d\lambda^{n-1}(y),$$

for Lipschitz boundaries, i.e., $\partial\Sigma^K$ and $\partial\Sigma$ are $C^{0,1}$-boundaries. Now we can estimate

$$\|T_2(g)\|^2_{H^{\frac{1}{2},2}(\partial\Sigma^K)} = \|T_2(g)\|^2_{L^2(\partial\Sigma^K)}$$
$$+ \sum_{i=1}^N \int_{B_1^{\mathbb{R}^{n-1}}(0)} \int_{B_1^{\mathbb{R}^{n-1}}(0)} \frac{|T_2(g)(\frac{\Psi_i(x,0)}{|\Psi_i(x,0)|^2}) - T_2(g)(\frac{\Psi_i(y,0)}{|\Psi_i(y,0)|^2})|^2}{|x-y|^n} d\lambda^{n-1}(x) d\lambda^{n-1}(y)$$
$$= \|g\|^2_{L^2(\partial\Sigma)}$$
$$+ \sum_{i=1}^N \int_{B_1^{\mathbb{R}^{n-1}}(0)} \int_{B_1^{\mathbb{R}^{n-1}}(0)} \frac{|g(\Psi_i(x,0)) - g(\Psi_i(y,0))|^2}{|x-y|^n} d\lambda^{n-1}(x) d\lambda^{n-1}(y)$$
$$= \|g\|^2_{H^{\frac{1}{2},2}(\partial\Sigma)},$$

3.2. TRANSFORMATIONS

for $g \in H^{\frac{1}{2},2}(\partial\Sigma)$. It is left to prove the continuity of

$$T_2 : \left(L^2(\partial\Sigma), \|\cdot\|_{H^{-\frac{1}{2},2}(\partial\Sigma)}\right) \to H^{-\frac{1}{2},2}(\partial\Sigma^K).$$

For $g \in L^2(\partial\Sigma) \subset H^{-\frac{1}{2},2}(\partial\Sigma)$ we use the definition of the integral on $\partial\Sigma$ and $\partial\Sigma^K$ from Definition 2.1.3. Furthermore we have

$$0 < C_{12} < J_i(y'), J_i^K(y') < C_{13} < \infty,$$

for $y' \in B_1^{\mathbb{R}^{n-1}}(0)$ and constants $0 < C_{12}, C_{13} < \infty$, see Lemma 2.1.4. Now we can start to estimate

$$|T_2(g)(h)| = \left|\int_{\partial\Sigma^K} T_2(g)(x)h(x)dH^{n-1}(x)\right| = \left|\int_{\partial\Sigma^K} g(\frac{x}{|x|^2})h(x)dH^{n-1}(x)\right|$$

$$\leq \sum_{i=1}^{N} \left|\int_{\mathbb{R}^{n-1}} w_i^K(\Psi_i^K(x',0))g(\frac{\Psi_i^K(x',0)}{|\Psi_i^K(x',0)|^2})h(\Psi_i^K(x',0))J_i^K(x')d\lambda^{n-1}(x')\right|$$

$$= \sum_{i=1}^{N} \left|\int_{\mathbb{R}^{n-1}} w_i(\Psi_i(x',0))g(\Psi_i(x',0))h(\frac{\Psi_i(x',0)}{|\Psi_i(x',0)|^2}) \cdot \underbrace{\frac{J_i^K(x')}{J_i(x')}}_{=:Q_i(\Psi_i^{-1}(x))} \cdot J_i(x')d\lambda^{n-1}(x')\right|$$

$$= \sum_{i=1}^{N} \left|\int_{\partial\Sigma} w_i g(x) h(\frac{x}{|x|^2}) Q_i(x) dH^{n-1}(x)\right| \leq \sum_{i=1}^{N} \|g\|_{H^{-\frac{1}{2},2}(\partial\Sigma)} \|w_i Q_i T_2(h)\|_{H^{\frac{1}{2},2}(\partial\Sigma)}.$$

Because $\partial\Sigma$ and consequently $\partial\Sigma$ are $C^{2,1}$-surfaces, we have $w_i Q_i \in H^{1,\infty}(\partial\Sigma)$. So using Lemma 2.3.6 and the the fact that T_2 is isometric between $H^{\frac{1}{2},2}(\partial\Sigma)$ and $H^{\frac{1}{2},2}(\partial\Sigma^K)$ we get

$$|T_2(g)(h)| \leq \|g\|_{H^{-\frac{1}{2},2}(\partial\Sigma)} c_9 \|h\|_{H^{\frac{1}{2},2}(\partial\Sigma^K)},$$

for all $g \in L^2(\partial\Sigma)$ and $h \in H^{\frac{1}{2},2}(\partial\Sigma^K)$. Using the BLT Theorem, see Lemma 2.4.5, again this proves the assertion. $\square$

Closing this section, we give the definitions of the transformations T_3 and T_4 as well as their most important properties in the final lemma of this subsection.

Lemma 3.2.14. Let Σ be an outer $C^{2,1}$-domain. We define the operators T_3 and T_4 via

$$T_3 : H^{2,\infty}(\partial\Sigma) \to H^{2,\infty}(\partial\Sigma^K),$$
$$(T_3(\underline{a}))(x) := |x|^n \cdot \left(\underline{a}(\frac{x}{|x|^2}) - 2\left\langle \underline{a}(\frac{x}{|x|^2}), \underline{e}_x \right\rangle \underline{e}_x\right), \quad x \in \partial\Sigma^K \quad (3.26)$$

for $\underline{a} \in H^{2,\infty}(\partial\Sigma)$ and
$$T_4 : H^{1,\infty}(\partial\Sigma) \to H^{1,\infty}(\partial\Sigma^K),$$
$$(T_4(b))(x) := |x|^{n-2} \cdot \left(b(\frac{x}{|x|^2}) + (2-n)\left\langle \underline{a}(\frac{x}{|x|^2}), x \right\rangle \right), \quad x \in \partial\Sigma^K \quad (3.27)$$
for $b \in H^{1,\infty}(\partial\Sigma)$,

where $\underline{e}_x$ denotes the unit vector in direction x and $\underline{a} \in H^{2,\infty}(\partial\Sigma)$ for T_4. Furthermore we have
$$T_3 : H^{1,\infty}(\partial\Sigma) \to H^{1,\infty}(\partial\Sigma^K),$$
$$T_4 : L^{\infty}(\partial\Sigma) \to L^{\infty}(\partial\Sigma^K),$$

if Σ is an outer $C^{1,1}$-domain and $\underline{a} \in H^{1,\infty}(\partial\Sigma)$ for T_4. All these operators are well defined and give the relation formulated by equations (3.21) and (3.22).

Proof. It can be verified easily using the product rule for Sobolev functions that the operators are well defined. In order to verify the relation given by equations (3.21) and (3.22), we compute the gradient of $K(u)$ for $u \in C^1(\overline{\Sigma})$. We have $K(u) \in C^1(\overline{\Sigma^K})$ and we have

$$\begin{aligned}
&\nabla((K(u))(x)) \\
&= \nabla(\frac{1}{|x|^{n-2}} u(\frac{x}{|x|^2})) \\
&= \nabla(\frac{1}{|x|^{n-2}}) u(\frac{x}{|x|^2}) + \frac{1}{|x|^{n-2}} \nabla(u(\frac{x}{|x|^2})) \\
&= \frac{2-n}{|x|^n} u(\frac{x}{|x|^2}) x + \frac{1}{|x|^{n-2}} (\nabla(u)(\frac{x}{|x|^2})) \circ \left(\partial_1 \frac{x}{|x|^2}, \ldots, \partial_n \frac{x}{|x|^2} \right) \\
&= \frac{2-n}{|x|^n} u(\frac{x}{|x|^2}) x - \frac{2}{|x|^{n+2}} (\nabla(u)(\frac{x}{|x|^2})) \circ (x_i x_j)_{i,j=1,\ldots,n} + \frac{1}{|x|^n} \nabla u(\frac{x}{|x|^2}) \\
&= \frac{2-n}{|x|^n} u(\frac{x}{|x|^2}) x + \frac{1}{|x|^n} \nabla u(\frac{x}{|x|^2}) \\
&\quad - \frac{2}{|x|^{n+2}} \partial_1 u(\frac{x}{|x|^2}) x_1 x - \ldots - \frac{2}{|x|^{n+2}} \partial_n u(\frac{x}{|x|^2}) x_n x \\
&= \frac{2-n}{|x|^n} u(\frac{x}{|x|^2}) x + \frac{1}{|x|^n} \nabla u(\frac{x}{|x|^2}) - \frac{2}{|x|^{n+2}} \left\langle \nabla u(\frac{x}{|x|^2}), x \right\rangle x.
\end{aligned}$$

for all $x \in \partial\Sigma$. $\circ$ denotes the Matrix multiplication. Plugging this into the boundary condition on $\partial\Sigma$ we get

$$\begin{aligned}
&\langle \underline{a}(x), \nabla K(u)(x) \rangle + b(x) \cdot K(u)(x) \\
&= \langle \underline{a}(x), x \rangle \frac{2-n}{|x|^n} u(\frac{x}{|x|^2}) + \frac{1}{|x|^n} \left\langle \underline{a}(x), \nabla u(\frac{x}{|x|^2}) \right\rangle
\end{aligned}$$

$$-\frac{2}{|x|^{n+2}}\left\langle x, \nabla u(\frac{x}{|x|^2})\right\rangle \langle \underline{a}(x), x\rangle + \frac{b(x)}{|x|^{n-2}}u(\frac{x}{|x|^2})$$
$$= \left\langle \left(\frac{1}{|x|^n}\underline{a}(x) - \frac{2}{|x|^{n+2}}\langle \underline{a}(x), x\rangle x\right) \cdot \nabla u(\frac{x}{|x|^2})\right\rangle$$
$$+ \left(\frac{b(x)}{|x|^{n-2}} + \langle \underline{a}(x), x\rangle \frac{2-n}{|x|^n}\right) u(\frac{x}{|x|^2})$$
$$= \langle T_3(\underline{a})(\frac{x}{|x|^2}), \nabla u(\frac{x}{|x|^2})\rangle + T_4(b)(\frac{x}{|x|^2})u(\frac{x}{|x|^2})$$

for all $x \in \partial \Sigma$. If we now use equation (3.21) together with (3.23) we will get equation (3.22) and the proof is done. $\square$

3.3 The Outer Oblique Boundary Problem of Potential Theory

This is the main section of Chapter 3. It presents a solution operator for weak solutions to the outer oblique boundary problem for the Poisson equation. In the first subsection we provide an existence result. The weak solution to the outer problem will be the Kelvin transformation $K(v)$ of the weak solution to the corresponding regular inner problem v. Next we state a regularization result for the weak solution. Thus the connection between the weak solution and the original problem is established. Then we discuss the transformed regularity condition on the oblique vector field. After we implemented stochastic inhomogeneities and provided a Ritz-Galerkin approximation, we show the applicability to problems from Geomathematics with some examples in the last subsection.

3.3.1 Weak Solutions to the Outer Problem

In this subsection we want apply the solution operator of the inner regular problem in order to get a weak solution of the outer problem. Therefore we will use a combination of all the operators defined in the previous sections and subsections. In order to avoid confusion we denote the normal vector of $\partial \Sigma$ by ν and the normal vector of $\partial \Sigma^K$ by ν^K. With the same reasoning as for the inner setting, we assume Σ to be at least an outer $C^{1,1}$-domain if not stated otherwise. We start with the classical formulation of the outer oblique boundary problem for the Poisson equation in the following definition.

Definition 3.3.1. Let Σ be an outer $C^{1,1}$-domain, $f \in C^0(\Sigma)$, $b, g \in C^0(\partial\Sigma)$ and $\underline{a} \in C^0(\partial\Sigma; \mathbb{R}^n)$ be given. A function $u \in C^2(\Sigma) \cap C^1(\overline{\Sigma})$ such that

$$\Delta u(x) = f(x), \quad \text{for all } x \in \Sigma, \tag{3.28}$$

$$\langle \underline{a}(x) \cdot \nabla u(x) \rangle + b \cdot u(x) = g(x), \quad \text{for all } x \in \partial\Sigma, \tag{3.29}$$

is called *classical solution* of the *outer oblique boundary problem for the Poisson equation*.

Now we state the main result of this chapter.

Theorem 3.3.2. *Let Σ be an outer $C^{1,1}$-domain. Furthermore let $\underline{a} \in H^{1,\infty}(\partial\Sigma; \mathbb{R}^n)$, $b \in L^\infty(\partial\Sigma)$, $g \in H^{-\frac{1}{2},2}(\partial\Sigma)$ and $f \in \left(H^{1,2}_{|x|^2,|x|^3}(\Sigma)\right)'$, such that*

$$\left|\langle (T_3(\underline{a}))(y), \nu^K(y) \rangle\right| > C_{14} > 0, \tag{3.30}$$

$$\operatorname{ess\,inf}_{\partial \Sigma^K} \left\{ \frac{T_4(b)}{\langle T_3(\underline{a}), \nu^K \rangle} - \frac{1}{2} \operatorname{div}_{\partial \Sigma^K} \left(\frac{T_3(\underline{a})}{\langle T_3(\underline{a}), \nu^K \rangle} - \nu^K \right) \right\} > 0, \tag{3.31}$$

for all $y \in \partial\Sigma^K$, where $0 < C_{14} < \infty$. Then we define

$$u := S^{\text{out}}_{\underline{a},b}(f, g) := K\left(S^{\text{in}}_{T_3(\underline{a}), T_4(b)}(T_1(f), T_2(g))\right), \tag{3.32}$$

as the weak solution to the outer oblique boundary problem for the Poisson equation from Definition 3.3.1. $S^{\text{out}}_{\underline{a},b}$ is injective and we have for a constant $0 < C_{15} < \infty$

$$\|u\|_{H^{1,2}_{\frac{1}{|x|^2}, \frac{1}{|x|}}(\Sigma)} \leq C_{15} \left(\|f\|_{\left(H^{1,2}_{|x|^2,|x|^3}(\Sigma)\right)'} + \|g\|_{H^{-\frac{1}{2},2}(\partial\Sigma)} \right). \tag{3.33}$$

Proof. Lemma 3.2.3 yields Σ^K to be a bounded $C^{1,1}$-domain. Furthermore, because of Lemma 3.2.14 and condition (3.30) we have that $T_3(\underline{a}) \in H^{1,\infty}(\partial\Sigma^K; \mathbb{R}^n)$ and condition (3.1) is fulfilled. Additionally $T_4(b) \in L^\infty(\partial\Sigma)$. So $S^{\text{in}}_{T_3(\underline{a}), T_4(b)}$ is well defined, see Remark 3.1.4. Injectivity follows by the injectivity of the operators K, T_1, T_2 and $S^{\text{in}}_{T_3(\underline{a}), T_4(b)}$. Lemma 3.2.11 and Lemma 3.2.13 yields $T_1(f) \in \left(H^{1,2}_{|x|^2,|x|^3}(\Sigma^K)\right)'$ and $T_2(g) \in H^{-\frac{1}{2},2}(\partial\Sigma^K)$. Consequently we can apply the solution operator for the regular inner problem to get a weak solution $v \in H^{1,2}(\Sigma^K)$ by

$$v := S^{\text{in}}_{T_3(\underline{a}), T_4(b)}(T_1(f), T_2(g)),$$

see Theorem 3.1.3. Using Lemma 3.2.7, we get $u := K(v) \in H^{1,2}_{\frac{1}{|x|^2}, \frac{1}{|x|}}(\Sigma)$. Finally, we use the continuity estimates from Lemmata 3.1.3, 3.2.7, 3.2.11 and 3.2.13, to obtain

$$\|u\|_{H^{1,2}_{\frac{1}{|x|^2}, \frac{1}{|x|}}(\Sigma)} \leq C_2 C_8 \max\{C_{10}, C_{11}\} \left(\|f\|_{\left(H^{1,2}_{|x|^2,|x|^3}(\Sigma)\right)'} + \|g\|_{H^{-\frac{1}{2},2}(\partial\Sigma)} \right).$$

$\square$

3.3. THE OUTER OBLIQUE BOUNDARY PROBLEM OF POTENTIAL THEORY

We will need the following lemma in order to prove a regularization result for the outer problem.

Lemma 3.3.3. Let Σ be an outer $C^{1,1}$-domain and $u \in H^{2,2}(\Sigma^K)$ be given. Then we have $K(u) \in H^{2,2}_{\frac{1}{|x|^2}, \frac{1}{|x|}, 1}(\Sigma)$ and

$$\|K(u)\|_{H^{2,2}_{\frac{1}{|x|^2}, \frac{1}{|x|}, 1}(\Sigma)} \leq C_{16} \|u\|_{H^{2,2}(\Sigma^K)},$$

for a constant $0 < C_{16} < \infty$ independent of u, i.e., $K : H^{2,2}(\Sigma^K) \to H^{2,2}_{\frac{1}{|x|^2}, \frac{1}{|x|}, 1}(\Sigma)$ defines a continuous operator.

Proof. Due to Lemma 3.2.7 we only have to estimate $\|\partial_j \partial_i K(u)\|_{L^2(\Sigma)}$ for $1 \leq i, j \leq n$. But this can be done easily by using the techniques as in the proof of Lemma 3.2.7. We have

$$\|\partial_j \partial_i \mathrm{KT}_F(u)\|^2_{L^2(\Sigma)}$$

$$= \int_\Sigma \left(\partial_j \left((2-n) \frac{x_i}{|x|^n} u(\frac{x}{|x|^2}) \right. \right.$$
$$\left. \left. + (\partial_i u)(\frac{x}{|x|^2}) \frac{1}{|x|^n} - 2 \sum_{k=1}^n (\partial_k u)(\frac{x}{|x|^2}) \frac{x_i x_k}{|x|^{n+2}} \right) \right)^2 d\lambda^n(x)$$

$$= \int_\Sigma \left((2-n) x_i u(\frac{x}{|x|^2}) \partial_j (\frac{1}{|x|^n}) + \right.$$
$$(2-n) \frac{x_i}{|x|^n} \partial_j (u(\frac{x}{|x|^2})) + \partial_j ((\partial_i u)(\frac{x}{|x|^2})) \frac{1}{|x|^n} + (\partial_i u)(\frac{x}{|x|^2}) \partial_j (\frac{1}{|x|^n})$$
$$\left. -2 \sum_{k=1}^n \partial_j ((\partial_k u)(\frac{x}{|x|^2})) \frac{x_i x_k}{|x|^{n+2}} - 2 \sum_{k=1}^n (\partial_k u)(\frac{x}{|x|^2}) \partial_j (\frac{x_i x_k}{|x|^{n+2}}) \right)^2 d\lambda^n(x)$$

$$= \int_\Sigma \left((n-2) n \frac{x_i x_j}{|x|^{n+2}} u(\frac{x}{|x|^2}) + \right.$$
$$(\partial_j u)(\frac{x}{|x|^2}) \frac{x_i(2-n)}{|x|^{n+2}} + \sum_{l=1}^n (\partial_l u)(\frac{x}{|x|^2}) \frac{(2-n) x_i x_j x_l}{|x|^{n+4}} + (\partial_j \partial_i u)(\frac{x}{|x|^2})) \frac{1}{|x|^{n+2}}$$
$$+ \sum_{l=1}^n (\partial_l \partial_i u)(\frac{x}{|x|^2}) \frac{x_j x_l}{|x|^{n+4}} - \frac{n x_j}{|x|^{n+2}} (\partial_i u)(\frac{x}{|x|^2}) + \delta_{ij} \cdot (2-n) \frac{1}{|x|^n} u(\frac{x}{|x|^2})$$
$$-2\delta_{ij} \sum_{k=1}^n (\partial_k u)(\frac{x}{|x|^2}) \frac{x_k}{|x|^{n+2}} - 2\delta_{jk} (\partial_k u)(\frac{x}{|x|^2}) \frac{x_i}{|x|^{n+2}}$$
$$\left. -2 \sum_{k=1}^n (\partial_j \partial_k u)(\frac{x}{|x|^2})) \frac{x_i x_k}{|x|^{n+4}} \right.$$

$$
\begin{aligned}
&\left. +2\sum_{k=1}^{n}\sum_{l=1}^{n}(\partial_l\partial_k u)(\frac{x}{|x|^2}))\frac{x_l x_i x_k}{|x|^{n+4}} - 2\sum_{k=1}^{n}(\partial_k u)(\frac{x}{|x|^2})\frac{x_j x_i x_k}{|x|^{n+4}}\right)^2 d\lambda^n(x) \\
\leq\ & C_{17}\int_\Sigma \frac{1}{|x|^{2n}}\left(u^2(\frac{x}{|x|^2}) + \sum_{i=1}^{n} u(\frac{x}{|x|^2})(\partial_i u)(\frac{x}{|x|^2}) + \sum_{i=1}^{n}\sum_{j=1}^{n} u(\frac{x}{|x|^2})(\partial_i\partial_j u)(\frac{x}{|x|^2})\right. \\
&+ \sum_{i=1}^{n}\sum_{j=1}^{n}(\partial_i u)(\frac{x}{|x|^2})(\partial_j u)(\frac{x}{|x|^2}) \\
&+ \sum_{i=1}^{n}\sum_{j=1}^{n}\sum_{k=1}^{n}(\partial_i u)(\frac{x}{|x|^2})(\partial_j\partial_k u)(\frac{x}{|x|^2}) \\
&\left. + \sum_{i=1}^{n}\sum_{j=1}^{n}\sum_{k=1}^{n}\sum_{l=1}^{n}(\partial_i\partial_j u)(\frac{x}{|x|^2})(\partial_k\partial_l u)(\frac{x}{|x|^2})\right) d\lambda^n(x) \\
=\ & C_{17}\int_{\Sigma^K} |y|^{2n}\left(u^2(y) + \sum_{i=1}^{n} u(y)(\partial_i u)(y) + \sum_{i=1}^{n}\sum_{j=1}^{n} u(y)(\partial_i\partial_j u)(y)\right. \\
&+ \sum_{i=1}^{n}\sum_{j=1}^{n}(\partial_i u)(y)(\partial_j u)(y) \\
&+ \sum_{i=1}^{n}\sum_{j=1}^{n}\sum_{k=1}^{n}(\partial_i u)(y)(\partial_j\partial_k u)(y) \\
&\left. + \sum_{i=1}^{n}\sum_{j=1}^{n}\sum_{k=1}^{n}\sum_{l=1}^{n}(\partial_i\partial_j u)(y)(\partial_k\partial_l u)(y)\right) |\mathrm{Det}(KT_D(y))|d\lambda^n(y) \\
\leq\ & C_{18}\int_{\Sigma^K} |y|^{2n}\left(u^2(y) + \sum_{i=1}^{n} u(y)(\partial_i u)(y) + \sum_{i=1}^{n}\sum_{j=1}^{n} u(y)(\partial_i\partial_j u)(y)\right. \\
&+ \sum_{i=1}^{n}\sum_{j=1}^{n}(\partial_i u)(y)(\partial_j u)(y) \\
&+ \sum_{i=1}^{n}\sum_{j=1}^{n}\sum_{k=1}^{n}(\partial_i u)(y)(\partial_j\partial_k u)(y) \\
&\left. + \sum_{i=1}^{n}\sum_{j=1}^{n}\sum_{k=1}^{n}\sum_{l=1}^{n}(\partial_i\partial_j u)(y)(\partial_k\partial_l u)(y)\right) |y|^{-2n}d\lambda^n(y) \\
\leq\ & C_{16}\|u\|^2_{H^{2,2}(\Sigma^K)}.
\end{aligned}
$$

Now the proof is finished. □

Now we are able to state the following regularization result, based on the regularization result for the inner problem, see Theorem 3.1.6. The following Theorem shows, that the weak solution, defined by Theorem 3.3.2, is really related to the outer problem, given in Definition 3.3.1.

Theorem 3.3.4. *Let Σ be an outer $C^{2,1}$-domain, $\underline{a} \in H^{2,\infty}(\partial\Sigma; \mathbb{R}^n)$, $b \in H^{1,\infty}(\partial\Sigma)$ such that (3.30) and (3.31) holds. If $f \in L^2_{|x|^2}(\Sigma)$ and $g \in H^{\frac{1}{2},2}(\Sigma)$ then we have $u \in H^{2,2}_{\frac{1}{|x|^2},\frac{1}{|x|},1}(\Sigma)$, for u provided by Theorem 3.3.2, and*

$$\Delta u = f, \qquad (3.34)$$

$$\langle \underline{a}, \nabla u\rangle + bu = g, \qquad (3.35)$$

almost everywhere on Σ and $\partial\Sigma$, respectively. Furthermore we have an a priori estimate

$$\|u\|_{H^{2,2}_{\frac{1}{|x|^2},\frac{1}{|x|},1}(\Sigma)} \leq C_{19}\Big(\|f\|_{L^2_{|x|^2}(\Sigma)} + \|g\|_{H^{\frac{1}{2},2}(\partial\Sigma)}\Big), \qquad (3.36)$$

with a constant $0 < C_{19} < \infty$. Such a solution we call strong solution to the outer oblique boundary problem for the Poisson equation.

Proof. Using Lemmata 3.2.3, 3.2.9, 3.2.13 and 3.2.14, we find the requirement of Theorem 3.1.6 to be fulfilled. Consequently we have $v \in H^{2,2}(\Sigma^K)$ and

$$\Delta v(x) = T_1(f)(x), \quad \text{for almost all } x \in \Sigma^K,$$
$$\langle T_3(\underline{a})(x), \nabla v(x)\rangle + T_4(b)(x) \cdot v(x) = T_2(g)(x), \quad \text{for almost all } x \in \partial\Sigma^K.$$

Furthermore, this yields $u \in H^{2,2}_{\frac{1}{|x|^2},\frac{1}{|x|},1}(\Sigma)$, by Lemma 3.3.3, and

$$\Delta u(y) = f(y), \quad \text{for almost all } y \in \Sigma,$$
$$\langle \underline{a}(y), \nabla u(y)\rangle + b(y) \cdot u(y) = g(y), \quad \text{for almost all } y \in \partial\Sigma,$$

by the choice of the transformations. $\square$

As a consequence we have that if the data in Theorem 3.3.4 fulfills the requirements of a classical solution, see e.g. [Mir70, Section 23], the solution u provided by Theorem 3.3.2 coincides with this solution. We close the subsection with a final remark.

Remark 3.3.5. It would also be possible to derive a weak formulation and to prove that the weak solution, provided by Theorem 3.3.2, fulfills it. This would require additional regularity assumptions on $\underline{a}$, b and Σ. Additionally, we would have to forbid tangential directions for the oblique vector field $\underline{a}$ and we would end up with a regular outer oblique boundary problem. Since we would still have to use the Kelvin transformation, there is no advantage in such a formulation at this point.

3.3.2 The Condition on the Oblique Vector Field

Analogously to the regular inner problem, we have condition (3.31), which is a transformed version of (3.5) and gives a relation between $\underline{a}$ and b, depending on the geometry of the surface $\partial \Sigma$. Moreover condition (3.30) is a transformed version of (3.1) and gives the non admissible direction for the oblique vector field $\underline{a}$. For the regular inner problem, (3.1) states the tangential directions as non admissible for the oblique vector field. For the outer problem the direction depends as well on the direction of the normal vector $\nu(y)$ at the point $y \in \partial \Sigma$ as on y itself. In this subsection we will investigate this dependency in detail. Using the definition of T_3 we get

$$\left| \left\langle |x|^n \left(\underline{a}(\frac{x}{|x|^2}) - 2 \left\langle \underline{a}(\frac{x}{|x|^2}), \underline{e}_x \right\rangle \underline{e}_x \right), \nu^K(x) \right\rangle \right| > C_{14} > 0,$$

for almost all $x \in \partial \Sigma^K$. This is equivalent to

$$\min \left(|y|^n \Big| y \in \partial \Sigma^K \right) \left| \left\langle \underline{a}(\frac{x}{|x|^2}), \nu^K(x) \right\rangle - 2 \left\langle \underline{a}(\frac{x}{|x|^2}), \underline{e}_x \right\rangle \langle \underline{e}_x, \nu^K(x) \rangle \right| > C_{20} > 0,$$

for all $x \in \partial \Sigma$ and $0 < C_{20} < \infty$. We use the formula

$$\frac{\langle y, z \rangle}{|y| \cdot |z|} =: \cos(\angle_{y,z}),$$

for vectors in $\mathbb{R}^n$, where $\angle_{y,z}$ denotes the angle $0 \leq \angle_{y,z} \leq \pi$ between y and z. Now, we can rewrite condition (3.30) in the equivalent form

$$\left| |\underline{a}(\frac{x}{|x|^2})| \cdot \cos(\angle_{\underline{a}(x), \nu^K}) - 2 \cdot |\underline{a}(\frac{x}{|x|^2})| \cdot \cos(\angle_{\underline{a}(x), \underline{e}_x}) \cdot \cos(\angle_{\underline{e}_x, \nu^K}) \right| > C_{21} > 0,$$

and finally

$$\left| \cos(\angle_{\underline{a}(x), \nu^K(x)}) - 2 \cdot \cos(\angle_{\underline{a}(x), \underline{e}_x}) \cdot \cos(\angle_{\underline{e}_x, \nu^K(x)}) \right| > C_{22} > 0, \qquad (3.37)$$

3.3. THE OUTER OBLIQUE BOUNDARY PROBLEM OF POTENTIAL THEORY

for all $x \in \partial \Sigma^K$ and constant $0 < C_{21}, C_{22} < \infty$ independent of x. This equation only states a condition on the direction of $\underline{a}$, not on its modulus. Going to $\mathbb{R}^2$ we can explicitly compute the non admissible direction resulting from condition (3.30). We can write

$$\cos(\angle_{\underline{a}(x),\nu^K(x)}) = \cos(\angle_{\underline{a}(x),\underline{e}_x} \pm \angle_{\underline{e}_x,\nu^K(x)})$$
$$= \cos(\angle_{\underline{a}(x),\underline{e}_x}) \cdot \cos(\angle_{\underline{e}_x,\nu^K(x)}) \mp \sin(\angle_{\underline{a}(x),\underline{e}_x}) \cdot \sin(\angle_{\underline{e}_x,\nu^K(x)}),$$

using an addition theorem for trigonometric functions, where the sign of the angle depends on the geometric positions of the three vectors ν^K, $\underline{e}_x$ and $\underline{a}$, to each other. Plugging this into equation (3.37), the non admissible direction is described by

$$\cos(\angle_{\underline{a}(x),\underline{e}_x}) \cdot \cos(\angle_{\underline{e}_x,\nu^K(x)}) \mp \sin(\angle_{\underline{a}(x),\underline{e}_x}) \cdot \sin(\angle_{\underline{e}_x,\nu^K(x)}) = 2 \cdot \cos(\angle_{\underline{a}(x),\underline{e}_x}) \cdot \cos(\angle_{\underline{e}_x,\nu^K(x)}),$$

which leads to

$$\mp \sin(\angle_{\underline{a}(x),\underline{e}_x}) \cdot \sin(\angle_{\underline{e}_x,\nu^K(x)}) = \cos(\angle_{\underline{a}(x),\underline{e}_x}) \cdot \cos(\angle_{\underline{e}_x,\nu^K(x)}),$$

Fixing $\angle_{\underline{e}_x,\nu^K(x)}$ as a given geometric constant only depending on x, we define

$$C_{23}(x) := \cos(\angle_{\underline{e}_x,\nu^K(x)}),$$
$$C_{24}(x) := \sin(\angle_{\underline{e}_x,\nu^K(x)}).$$

So we can write the condition on $\underline{a}$ in the equivalent form

$$\mp C_{24}(x) \sin(\angle_{\underline{a}(x),\underline{e}_x}) = C_{23}(x) \cos(\angle_{\underline{a}(x),\underline{e}_x}).$$

Finally, the transformed non admissible direction is characterized by

$$\angle_{\underline{a}(x),\underline{e}_x} = \tan^{-1} \left| \frac{C_{23}(x)}{C_{24}(x)} \right|,$$

if $C_{24}(x) \neq 0$ and $\angle_{\underline{a}(x),\underline{e}_x} = \frac{\pi}{2}$ if $C_{24}(x) = 0$. Generally, transforming the problem to an inner setting transforms the conditions for the coefficients $\underline{a}$ and b. There are circumstances in which we have the same non admissible direction as for the inner problem, i.e., the tangential directions are non admissible. For example, this is the case if $\partial \Sigma$ is a sphere around the origin. In Figure 5 the situation for $\Sigma \subset \mathbb{R}^2$ is illustrated, the dashed line indicates the non admissible direction, which occurs because of the transformed regularity condition $\langle T_3(\underline{a}), \nu^K \rangle > C_{14} > 0$, see (3.30).

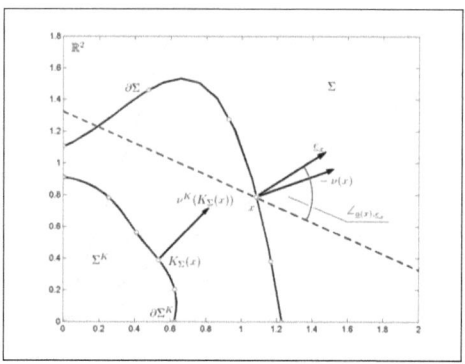

Figure 5: Non-admissible direction for the outer problem

3.3.3 Stochastic Inhomogeneities

In this subsection we implement stochastic inhomogeneities as well as stochastic weak solutions. For the inner problem stochastic inhomogeneities are already implemented, see [Ras05] or [GR06]. We will now provide them for the outer setting. We start by defining the spaces of stochastic functions. As already mentioned above, we denote in this subsection the sigma algebra by Σ, while space domains in $\mathbb{R}^n$ an their boundaries are denoted by Γ and $\partial \Gamma$, respectively. So, let Γ be an outer $C^{1,1}$-domain and (Ω, Σ, P) a probability space, arbitrary but fixed, such that $L^2(\Omega, P)$ is separable. We define

$$\left(H^{2,2}_{\frac{1}{|x|^2},\frac{1}{|x|},1}(\Gamma)\right)_\Omega := L^2(\Omega, P) \otimes H^{2,2}_{\frac{1}{|x|^2},\frac{1}{|x|},1}(\Gamma),$$
$$\left(H^{1,2}_{\frac{1}{|x|^2},\frac{1}{|x|}}(\Gamma)\right)_\Omega := L^2(\Omega, P) \otimes H^{1,2}_{\frac{1}{|x|^2},\frac{1}{|x|}}(\Gamma),$$
$$\left(L^2_{|x|^2}(\Gamma)\right)_\Omega := L^2(\Omega, P) \otimes L^2_{|x|^2}(\Gamma),$$
$$\left(H^{1,2}_{|x|^2,|x|^3}(\Gamma)\right)'_\Omega := L^2(\Omega, P) \otimes \left(H^{1,2}_{|x|^2,|x|^3}(\Gamma)\right)',$$
$$H^{\frac{1}{2},2}_\Omega(\partial \Gamma) := L^2(\Omega, P) \otimes H^{\frac{1}{2},2}(\partial \Gamma),$$
$$L^2_\Omega(\partial \Gamma) := L^2(\Omega, P) \otimes L^2(\partial \Gamma),$$
$$H^{-\frac{1}{2},2}_\Omega(\partial \Gamma) := L^2(\Omega, P) \otimes H^{-\frac{1}{2},2}(\partial \Gamma).$$

3.3. THE OUTER OBLIQUE BOUNDARY PROBLEM OF POTENTIAL THEORY

Because all spaces above are separable, see e.g. [Ada75], we can use the isomorphisms to Hilbert space valued random variables, see Lemma 2.4.3. We have the following main result of this section.

Theorem 3.3.6. *Let Γ be an outer $C^{1,1}$-domain. Furthermore let $\underline{a} \in H^{1,\infty}(\partial\Gamma;\mathbb{R}^n)$, $b \in L^{\infty}(\partial\Gamma)$, $g \in H_{\Omega}^{-\frac{1}{2},2}(\partial\Gamma)$ and $f \in \left(H_{|x|^2,|x|^3}^{1,2}(\Gamma)\right)'_{\Omega}$, such that (3.30) and (3.31) holds. Then we define*

$$u(\,\cdot\,,\omega) := S_{\underline{a},b}^{\text{out}}(f(\,\cdot\,,\omega), g(\,\cdot\,,\omega)), \tag{3.38}$$

for dP-almost all $\omega \in \Omega$. u is called stochastic weak solution to the outer oblique boundary problem for the Poisson equation. Furthermore we have for a constant $0 < C_{25} < \infty$

$$\|u\|_{\left(H_{\frac{1}{|x|^2},\frac{1}{|x|}}^{1,2}(\Gamma)\right)_{\Omega}} \leq C_{25} \left(\|f\|_{\left(H_{|x|^2,|x|^3}^{1,2}(\Gamma)\right)'_{\Omega}} + \|g\|_{H_{\Omega}^{-\frac{1}{2},2}(\partial\Gamma)}\right). \tag{3.39}$$

Proof. Using the isomorphisms stated in Lemma 2.4.3 we apply Theorem 3.3.2 and the proof is done. □

Finally, we have the following result for a stochastic strong solution.

Theorem 3.3.7. *Let Γ be an outer $C^{2,1}$-domain, $\underline{a} \in H^{2,\infty}(\partial\Gamma;\mathbb{R}^n)$, $b \in H^{1,\infty}(\partial\Gamma)$ such that (3.30) and (3.31) holds. If $f \in \left(L_{|x|^2}^{2}(\Gamma)\right)_{\Omega}$ and $g \in H_{\Omega}^{\frac{1}{2},2}(\Gamma)$ then we have $u \in \left(H_{\frac{1}{|x|^2},\frac{1}{|x|},1}^{2,2}(\Gamma)\right)_{\Omega}$, for u provided by Theorem 3.3.6, and*

$$\Delta u(x,\omega) = f(x,\omega), \tag{3.40}$$

$$\langle \underline{a}(y), \nabla u(y,\omega)\rangle + b(y)u(y,\omega) = g(y,\omega), \tag{3.41}$$

for λ^n-almost all $x \in \Gamma$, for H^{n-1}-almost all $y \in \partial\Gamma$ and for P-almost all $\omega \in \Omega$. Furthermore, we have an a priori estimate

$$\|u\|_{\left(H_{\frac{1}{|x|^2},\frac{1}{|x|},1}^{2,2}(\Gamma)\right)_{\Omega}} \leq C_{26}\left(\|f\|_{\left(L_{|x|^2}^2(\Gamma)\right)_{\Omega}} + \|g\|_{H_{\Omega}^{\frac{1}{2},2}(\partial\Gamma)}\right), \tag{3.42}$$

with a constant $0 < C_{26} < \infty$. Such a solution we call stochastic strong solution to the outer oblique boundary problem for the Poisson equation.

Proof. Using the isomorphisms stated in Lemma 2.4.3 we apply Theorem 3.3.4 and the proof is done. □

Remark 3.3.8. Alternatively we can use the stochastic solution operator $S_{\underline{a},b}^{\Omega,\mathrm{in}}(f,g)$ for the stochastic inner problem, provided by [GR06, Theorem 4.4.]. Therefore we have to define transformations for inhomogeneities and solution between the stochastic spaces. Then we define the stochastic weak solution to the outer problem by

$$u := K^\Omega \left(S_{T_3(\underline{a}),T_4(b)}^{\Omega,\mathrm{in}}(T_1^\Omega(f), T_2^\Omega(g)) \right)$$

and we end up with the same stochastic weak solution and the same results about its properties.

3.3.4 Ritz-Galerkin Approximation

In this subsection we provide the Ritz-Galerkin method which allows us to approximate the weak solution of the outer problem with help of a numerical computation. Therefore we use the approximation of the weak solution to the corresponding inner problem, provided by [GR06, Section 6] and [Alt02, Section 7.23]. Now assume Σ to be an outer $C^{1,1}$-domain. Furthermore let $\underline{a} \in H^{1,\infty}(\partial\Sigma;\mathbb{R}^n)$, $b \in L^\infty(\partial\Sigma)$, $g \in H^{-\frac{1}{2},2}(\partial\Sigma)$ and $f \in \left(H^{1,2}_{|x|^2,|x|^3}(\Sigma) \right)'$, such that condition (3.30) and condition (3.31) is fulfilled. We want to approximate the weak solution u to the outer oblique boundary problem, provided by Theorem 3.3.2. Let a and F be defined by

$$a(\eta, v) := -\sum_{i=1}^n {}_{H^{\frac{1}{2},2}(\partial\Sigma)}\left\langle \eta \frac{T_3(\underline{a})_i}{\langle T_3(\underline{a}),\nu^K\rangle} - \nu_i^K, (\nabla_{\partial\Sigma} v)_i \right\rangle_{H^{-\frac{1}{2},2}(\partial\Sigma)} -$$

$$\int_\Sigma (\nabla\eta, \nabla v)\, d\lambda^n - \int_{\partial\Sigma} \eta \frac{T_4(b)}{\langle T_3(\underline{a}),\nu^K\rangle} v\, dH^{n-1}$$

$$F(\eta) :=_{H^{\frac{1}{2},2}(\partial\Sigma)}\left\langle \eta, \frac{T_2(g)}{\langle T_3(\underline{a}),\nu^K\rangle} \right\rangle_{H^{-\frac{1}{2},2}(\partial\Sigma)} -_{H^{1,2}(\Sigma)}\langle \eta, T_1(f)\rangle_{(H^{1,2}(\Sigma))'}$$

for $\eta, v \in H^{1,2}(\Sigma^K)$.

Lemma 3.3.9. Let $(V_n)_{n\in\mathbb{N}}$ be a increasing sequence of finite dimensional subspaces of $H^{1,2}(\Sigma^K)$, i.e., $V_n \subset V_{n+1}$ such that $\overline{\bigcup_{n\in\mathbb{N}} V_n} = H^{1,2}(\Sigma^K)$. Then there exists for each $n \in \mathbb{N}$ an unique $v_n \in V_n$ with:

$$a(\eta, v_n) = F(\eta) \quad \text{for all } \eta \in V_n.$$

Proof. For the proof see [GR06, Lemma 6.1]. □

3.3. THE OUTER OBLIQUE BOUNDARY PROBLEM OF POTENTIAL THEORY

Lemma 3.3.10. Let $d := \dim(V_n)$ and $(\varphi_k)_{1 \leq k \leq d}$ be a basis of V_n. Then $v_n \in V_n$ from Lemma 3.3.9 has the following unique representation

$$v_n = \sum_{i=1}^{d} h_i \varphi_i,$$

where $(h_i)_{1 \leq i \leq d}$ is the solution of the linear system of equations given by

$$\sum_{i=1}^{d} a_{ji} h_i = F_j \quad 1 \leq j \leq d.$$

Here $a_{ji} := a(\varphi_j, \varphi_i)$ and $F_j := F(\varphi_j)$.

Proof. For the proof see [GR06, Lemma 6.2]. □

The following lemma from Céa proves that the sequence $(v_n)_{n \in \mathbb{N}}$ really approximates the solution v.

Lemma 3.3.11. Let v be the weak solution provided by Theorem 3.1.3 to the corresponding inner problem in Σ^K and $(v_n)_{n \in \mathbb{N}}$ taken from Lemma 3.3.9. Then:

$$\|v - v_n\|_{H^{1,2}(\Sigma)} \leq \frac{c_{11}}{c_{12}} \operatorname{dist}(v, V_n) \stackrel{n \to \infty}{\longrightarrow} 0. \tag{3.43}$$

Proof. For the proof of this lemma see [GR06, Lemma 6.3]. □

Consequently we have

Theorem 3.3.12. Let u be the weak solution provided by Theorem 3.3.2 to the outer problem and v, $(v_n)_{n \in \mathbb{N}}$ taken from Lemma 3.3.11, both corresponding to $\underline{a}$, b, g, f and Σ, given at the beginning of this subsection. Then:

$$\|u - K(v_n)\|_{H^{1,2}(\Sigma)} \leq C_8 \frac{c_{11}}{c_{12}} \operatorname{dist}(v, V_n) \stackrel{n \to \infty}{\longrightarrow} 0. \tag{3.44}$$

Proof. This follows using equation (3.43) from the previous lemma together with the continuity of the Kelvin transformation, see Lemma 3.2.7. The constants c_{11} and c_{12} are the constants from the Lax-Milgram Lemma, see Lemma 2.4.4, corresponding to a and F from the beginning of this subsection. □

Remark 3.3.13. An analogous approximation for the stochastic weak solution is available, using the results from Subsection 3.3.3, this subsection as well as [GR06, Section 6].

3.3.5 Geomathematical Applications and Examples

In this subsection we give examples for stochastic data. This may be used in geomathematical applications in order to model noise on measured values. In the following we give the examples for the outer problem. They are also suitable for the inner problem, see [GR06]. Again we denote the sigma algebra by Σ as usual, in order to avoid confusion. For the domains we replace Σ by Γ and $\partial\Sigma$ by $\partial\Gamma$.

Gaussian inhomogeneities

We choose the probability space (Ω, Σ, P), such that X_i, $1 \leq i \leq n_1$, are $P \otimes \lambda^n$-measurable and Y_j, $1 \leq j \leq n_2$, are $P \otimes H^{n-1}$-measurable with $X_i(\cdot, x)$, $x \in \Gamma$, and $Y_j(\cdot, x)$, $x \in \partial\Gamma$, Gaussian random variables with expectation value 0 and variance $f^2_{\sigma_i}(x)$ or variance $g^2_{\sigma_j}(x)$, respectively. Here $f_{\sigma_i} \in L^2_{|x|^2}(\Gamma)$ and $g_{\sigma_j} \in L^2(\partial\Gamma)$. We define:

$$f(\omega, x) := f_\mu(x) + \sum_{i=1}^{n_1} X_i(\omega, x), \qquad g(\omega, x) := g_\mu(x) + \sum_{j=1}^{n_2} Y_j(\omega, x),$$

where $f_\mu \in L^2_{|x|^2}(\Gamma)$ and $g_\mu \in L^2(\partial\Gamma)$. To use such kind of inhomogeneities we must show

$$f \in L^2(\Omega \times \Gamma, P \otimes |x|^4 \cdot \lambda^n) \quad \text{and} \quad g \in L^2(\Omega \times \partial\Gamma, P \otimes H^{n-1}).$$

Lemma 3.3.14. *For f_μ, $f_{\sigma_i} \in L^2_{|x|^2}(\Gamma)$, $1 \leq i \leq n_1$ one has that $f \in L^2(\Omega \times \Gamma, P \otimes |x|^4 \cdot \lambda^n)$.*

Proof. For the proof see [GR06, Proposition 5.1]. □

Lemma 3.3.15. *For g_μ, $g_{\sigma_j} \in L^2(\partial\Gamma)$, $1 \leq j \leq n_2$ one has that $g \in L^2(\Omega \times \partial\Gamma, P \otimes H^{n-1})$.*

Proof. For the proof see [GR06, Proposition 5.1]. □

Gauß-Markov model

Here we refer to [FM02] in which an application of the Example from the previous paragraph can be found. The authors use a random field

$$h(\omega, x) := H(x) + Z(\omega, x)$$

3.3. THE OUTER OBLIQUE BOUNDARY PROBLEM OF POTENTIAL THEORY

to model an observation noise, where $x \in \partial B_1(0) \subset \mathbb{R}^3$ and $\omega \in \Omega$ with (Ω, Σ, P) a probability space. Here one has that $Z(\cdot, x)$, $x \in \partial B_1(0)$, is a Gaussian random variable with expectation value 0 and variance $\sigma^2 > 0$. Additionally $H(x) \in L^2(\partial B_1(0))$ and the covariance is given by:

$$\operatorname{cov}(Z(\cdot, x_1), Z(\cdot, x_2)) = K(x_1, x_2),$$

where $K : \partial B_1(0) \times \partial B_1(0) \to \mathbb{R}$ is a suitable kernel.

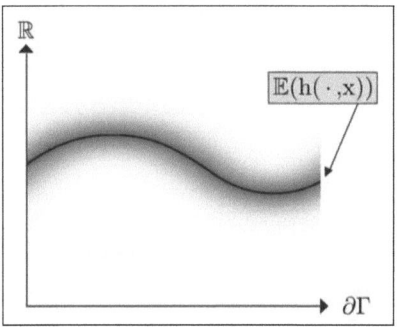

Figure 6: Data with Gaussian noise

Two geophysically relevant kernels are for example

$$K_1(x_1, x_2) := \frac{\sigma^2}{(M+1)^2} \sum_{n=1}^{M} \frac{2n+1}{4\pi} P_n((x_1 \cdot x_2)) \quad 0 \leq M < \infty,$$

$$K_2(x_1, x_2) := \frac{\sigma^2}{\exp(-c)} \exp(-c(x_1 \cdot x_2)).$$

P_n, $1 \leq n \leq M$, are the Legendre polynomials defined on $\mathbb{R}$. The noise model corresponding to the second kernel is called first degree Gauß–Markov model. If one chooses a $P \otimes H^{n-1}$-measurable random field Z, then h fulfills the requirements of Lemma 3.3.15. Existence of a corresponding probability measure P is provided in infinite dimensional Gaussian Analysis, see e.g. [For05].

Noise model for satellite data

In this paragraph we give another precise application, which can be found in [Bau04]. Here the authors are using stochastic inhomogeneities to implement a noise model for satellite data.

Therefore random fields of the form

$$h(\omega, x) := \sum_{i=1}^{m} h_i(x) Z_i(\omega)$$

are used, where $x \in \partial\Gamma \subset \mathbb{R}^3$ and $\omega \in \Omega$ with (Ω, Σ, P) a suitable probability space. Here $\partial\Gamma$ could be for example the earth's surface and we are searching for harmonic functions in the space outside the earth. Z_i are Gaussian random variables with expectation value 0 and variance $\sigma_i^2 > 0$ and h_i fulfilling the assumptions of Lemma 3.3.14, respectively Lemma 3.3.15. If one chooses (Ω, Σ, P) as $(\mathbb{R}^m, \mathbf{B}(\mathbb{R}), \gamma_{cov_{ij}}^{0,\sigma_i})$, where:

$$\gamma_{cov_{ij}}^{0,\sigma_i} := \frac{1}{\sqrt{(2\pi)^m det(A)}} e^{-\frac{1}{2}(\underline{y}, A^{-1}\underline{y})} d\lambda^m,$$
$$a_{ij} := \operatorname{cov}(Z_i, Z_j) \quad 1 \leq i, j \leq m,$$

one has a realization of Z_i as the projection on the i-th component in the separable space $L^2(\mathbb{R}^m, \gamma_{cov_{ij}}^{0,\sigma_i})$.

Chapter 4

Limit Formulae and Jump Relations

In this chapter we combine the modern theory of Sobolev spaces with the classical theory of limit formulae and jump relations of potential theory. Also other authors proved the convergence in Lebesgue spaces on $\partial\Sigma$ for integrable functions, see for example [Fic48], [Gün57] or [Ker80]. The achievement of this paper is the $L^2(\partial\Sigma)$ convergence for the weak derivatives of higher orders. Also the layer functions F are elements of Sobolev spaces and $\partial\Sigma$ is a two dimensional suitable smooth submanifold in $\mathbb{R}^3$, called $C^{m,\alpha}$-surface. We are considering the potential of the single layer, the potential of the double layer and their first order normal derivatives. Main tool is the convergence in $C^m(\partial\Sigma)$ which is proved with help of some results taken from [CK83], [FM04], [Kel67] and [Sch31b], together with a result from [Gün57]. Additionally, we need a result about the limit formulae in $L^2(\partial\Sigma)$, which can be found in [Ker80], and a reduction result which we found in [Mül51]. Moreover we prove the convergence in the Hölder spaces $C^{m,\beta}(\partial\Sigma)$. Finally we give an application of the limit formulae and jump relations. We generalize a density results from geomathematics, see [FK80], [FM03] and [FM04], based on the results proved before. Therefore we also prove the limit formula of U_1 for $F \in (H^{m,2}(\partial\Sigma))'$. During this chapter we always consider outer domains in $\mathbb{R}^3$, defined in Definition 2.1.1. Moreover, we assume Σ to be at least to be an outer C^2-domain if not stated otherwise.

4.1 Definition and Properties of the Layer Potentials

In this section we define the potential of the single layer, the potential of the double layer as well as their first order normal derivatives. This four potentials will be subject of this dissertation and their limit when approaching to the surface $\partial\Sigma$ will be investigated in several norms during the next sections. In this section we give the definition as well as some important properties of them.

Definition 4.1.1. Let $F \in C^0(\partial\Sigma)$, where Σ is an outer C^2-domain. The *potential of the single layer* on $\partial\Sigma$, denoted by $U_1[F]$ is defined by

$$U_1[F](x) := \int_{\partial\Sigma} F(y) \frac{1}{|x-y|} d\partial\Sigma(y)$$

for all $x \in \mathbb{R}^3 \backslash \partial\Sigma$. The *potential of the double layer* on $\partial\Sigma$, denoted by $U_2[F]$ is defined by

$$U_2[F](x) := \int_{\partial\Sigma} F(y) \frac{\partial}{\partial\nu(y)} \frac{1}{|x-y|} d\partial\Sigma(y)$$

for all $x \in \mathbb{R}^3 \backslash \partial\Sigma$. F is called *layer function*.

This two potentials have the following property.

Lemma 4.1.2. Let $F \in C^0(\partial\Sigma)$ where Σ is an outer C^2-domain. We have $U_1[F], U_2[F] \in C^\infty(\mathbb{R}^3 \backslash \partial\Sigma)$ and

$$\Delta U_1[F](x) = 0,$$
$$\Delta U_2[F](x) = 0,$$

for all $x \in \mathbb{R}^3 \backslash \partial\Sigma$. Furthermore, U_1 and U_2 are regular at infinity, i.e., $U_1(x) \to 0$ and $U_2(x) \to 0$ for $|x|$ tending to infinity.

Proof. Both, the proof of this lemma and the definition above can be found in [FM04]. □

Now we give the final definition of this section. Recall the definition of the normal vector field on $B_{r_0}(\partial\Sigma)$ in Lemma 2.1.6.

Definition 4.1.3. Let $F \in C^0(\partial\Sigma)$, where Σ is an outer C^2-domain. We define the first order normal derivative of $U_1[F]$, denoted by $\frac{\partial U_1}{\partial\nu}[F]$, and the first order normal derivative of $U_2[F]$, denoted by $\frac{\partial U_2}{\partial\nu}[F]$, via

$$\frac{\partial U_1}{\partial\nu}[F](x) := \langle \nu(x), \nabla U_1(x)[F] \rangle,$$
$$\frac{\partial U_2}{\partial\nu}[F](x) := \langle \nu(x), \nabla U_2(x)[F] \rangle,$$

for all $x \in B_{r_0}(\partial\Sigma) \backslash \partial\Sigma$.

4.2 Pointwise and Uniform Convergence

In this section we state the limit formulae of potential theory, pointwise and uniformly on $\partial\Sigma$. The most results presented in this section are well known from literature. In this cases we give references instead of proofs.

Definition 4.2.1. Let Σ be an outer C^2-domain, $F \in C^0(\partial\Sigma)$ and $F \in C^{1,\alpha}(\partial\Sigma)$, $0 < \alpha \leq 1$, for $\frac{\partial U_2}{\partial \nu}[F]$, respectively. For $x \in \partial\Sigma$ we define:

$$U_1[F](x) := \int_{\partial\Sigma} F(y) \frac{1}{|x-y|} d\partial\Sigma(y),$$

$$\frac{\partial U_1}{\partial \nu}[F](x) := \frac{\partial}{\partial \nu(x)} \int_{\partial\Sigma} F(y) \frac{1}{|x-y|} d\partial\Sigma(y) = \int_{\partial\Sigma} F(y) \frac{\partial}{\partial \nu(x)} \frac{1}{|x-y|} d\partial\Sigma(y),$$

$$U_2[F](x) := \int_{\partial\Sigma} F(y) \frac{\partial}{\partial \nu(y)} \frac{1}{|x-y|} d\partial\Sigma(y),$$

$$\frac{\partial U_2}{\partial \nu}[F](x) = \frac{\partial}{\partial \nu(x)} \int_{\partial\Sigma} F(y) \frac{\partial}{\partial \nu(y)} \frac{1}{|x-y|} d\partial\Sigma(y)$$

$$= -\int_{\partial\Sigma} \left\langle \nu(x), \left[\nabla_x \frac{1}{|x-y|}, \nabla_{\partial\Sigma} F(y) \times \nu(y)\right]\right\rangle d\partial\Sigma(y)$$

$$= \int_{\partial\Sigma} \frac{F(y) - F(x)}{|x-y|^3} \cdot \left(\langle \nu(y), \nu(x)\rangle - 3\langle y-x, \nu(y)\rangle \langle y-x, \nu(x)\rangle\right) d\partial\Sigma(y).$$

$\cdot \times \cdot$ denotes the vector product in $\mathbb{R}^3$. All integrals are well defined, at least as Cauchy principal value, see for example [FM04] or [Kel67]. The existence of U_1, U_2 and the normal derivative of U_1 on $\partial\Sigma$ can be found in [FM04] and [Gün57, Paragraph 5]. The existence of the normal derivative of U_2 on $\partial\Sigma$ and its given representations are proved in [CK83] and [Sch31b]. Now we state the limit formulae of potential theory.

Theorem 4.2.2. Let Σ be an outer C^2-domain, $F \in C^0(\partial\Sigma)$ and $F \in C^{1,\alpha}(\partial\Sigma)$, $0 < \alpha \leq 1$, for $\frac{\partial U_2}{\partial \nu}[F]$, respectively. The limit formulae of potential theory are given by

$$\lim_{\tau \to 0^+} U_1[F](x \pm \tau\nu(x)) = U_1[F](x), \quad \forall x \in \partial\Sigma,$$

$$\lim_{\tau \to 0^+} \frac{\partial U_1}{\partial \nu}[F](x \pm \tau\nu(x)) = \frac{\partial U_1}{\partial \nu}[F](x) \mp 2\pi F(x), \quad \forall x \in \partial\Sigma,$$

$$\lim_{\tau \to 0^+} U_2[F](x \pm \tau\nu(x)) = U_2[F](x) \pm 2\pi F(x), \quad \forall x \in \partial\Sigma,$$

$$\lim_{\tau \to 0^+} \frac{\partial U_2}{\partial \nu}[F](x \pm \tau\nu(x)) = \frac{\partial U_2}{\partial \nu}[F](x), \quad \forall x \in \partial\Sigma.$$

The convergence is even uniformly in $x \in \partial\Sigma$.

Proof. For all formulae except of the last one, we refer to e.g. [CK83]. We only have to prove the lemma for the last formula. Due to results from [CK83, Theorem 2.23] we obtain all terms to be continuous on $\partial\Sigma$ for all $\tau \in (0, \tau_0]$. Additionally the authors prove:

$$\lim_{\tau \to 0^+} \nabla U_2[F](x \pm \tau\nu(x)) = -\int_{\partial\Sigma} \left[\nabla_x \frac{1}{|x-y|}, [\nabla_{\partial\Sigma}F(y), \nu(y)] \right] dH^2(y) \pm 2\pi \nabla_{\partial\Sigma}F(x),$$

uniformly in $x \in \partial\Sigma$. In turn, results from [Sch31b] yield

$$\lim_{\tau \to 0^+} \frac{\partial U_2}{\partial \nu}[F](x + \tau\nu(x)) = \frac{\partial U_2}{\partial \nu}[F](x), \quad \forall x \in \partial\Sigma,$$

on $\partial\Sigma$. This gives

$$-\left\langle \nu(x), \int_{\partial\Sigma} \left[\nabla_x \frac{1}{|x-y|}, [\nabla_{\partial\Sigma}F(y), \nu(y)] \right] dH^2(y) \right\rangle = \frac{\partial U_2}{\partial \nu}[F](x),$$

for all $x \in \partial\Sigma$, and we have proved

$$\lim_{\tau \to 0^+} \frac{\partial U_2}{\partial \nu}[F](x \pm \tau\nu(x)) = \frac{\partial U_2}{\partial \nu}[F](x),$$

uniformly for all $x \in \partial\Sigma$. Thus we are done. $\square$

Because we want to prove these limit formulae for several norms, we define for each of them a family of operators in the following lemma.

Lemma 4.2.3. Let Σ be an outer C^2-domain. Furthermore, let $F \in C^0(\partial\Sigma)$ and $F \in C^{1,\alpha}(\partial\Sigma)$, $\alpha \in (0,1]$, for $\frac{\partial U_2}{\partial \nu}[F]$, respectively. We define the following families of operators $\left(L_i^{\pm\tau}[F]\right)_{\tau \in (0,\tau_0]}$, $i = 1, 2, 3, 4$, by

$$L_1^{\pm\tau}[F] := U_1[F](x \pm \tau\nu(x)) - U_1(x),$$

$$L_2^{\pm\tau}[F] := \frac{\partial U_1}{\partial \nu}[F](x \pm \tau\nu(x)) - \frac{\partial U_1}{\partial \nu}[F](x) \pm 2\pi F(x),$$

$$L_3^{\pm\tau}[F] := U_2[F](x \pm \tau\nu(x)) - U_2[F](x) \mp 2\pi F(x),$$

$$L_4^{\pm\tau}[F] := \frac{\partial U_2}{\partial \nu}[F](x \pm \tau\nu(x)) - \frac{\partial U_2}{\partial \nu}[F](x),$$

for all $x \in \partial\Sigma$ and $\tau \in (0, \tau_0]$. We have

$$\lim_{\tau \to 0^+} L_i^{\pm\tau}[F](x) = 0, \tag{4.1}$$

4.2. POINTWISE AND UNIFORM CONVERGENCE

for all $x \in \partial\Sigma$ and $i = 1, 2, 3, 4$. Furthermore, we have $L_i^{\pm\tau}[F] \in C^0(\partial\Sigma)$ for $i = 1, 2, 3, 4$ and all $\tau \in (0, \tau_0]$ with

$$\lim_{\tau \to 0^+} \|L_i^{\pm\tau}[F]\|_{C^0(\partial\Sigma)} = 0, \qquad (4.2)$$

for $i = 1, 2, 3, 4$.

Proof. This follows by Theorem 4.2.2. □

We close the section with a remark.

Remark 4.2.4. It is easy to see, that the potentials perform certain jumps when either approaching from the inner space or from the outer space. We can easily conclude the so called jump relations from the corresponding limit formulae

$$\lim_{\tau \to 0^+} U_1[F](x + \tau\nu(x)) - U_1[F](x - \tau\nu(x)) = 0,$$
$$\lim_{\tau \to 0^+} \frac{\partial U_1}{\partial \nu}[F](x + \tau\nu(x)) - \frac{\partial U_1}{\partial \nu}[F](x - \tau\nu(x)) = -4\pi F(x),$$
$$\lim_{\tau \to 0^+} U_2[F](x + \tau\nu(x)) - U_2[F](x - \tau\nu(x)) = 4\pi F(x),$$
$$\lim_{\tau \to 0^+} \frac{\partial U_2}{\partial \nu}[F](x + \tau\nu(x)) - \frac{\partial U_2}{\partial \nu}[F](x - \tau\nu(x)) = 0,$$

for all $x \in \partial\Sigma$. In Figure 7 the jump relation for $U_2[F]$ is illustrated. $\frac{\partial U_1}{\partial \nu}[F]$ shows a similar behavior.

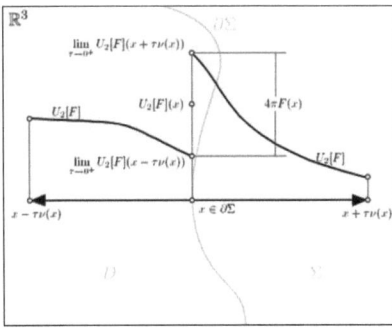

Figure 7: Boundary behavior of $U_2[F]$

Since we are interested in the limit formulae, we will omit the jump relations to simplify the exposition appreciably. They can be obtained directly without any additional considerations or conditions with help of the triangle inequality. Further results about the jump relations can be found in e.g. [FM04], [Ker80] and [Mül69].

4.3 Limit Formulae in $C^m(\partial\Sigma)$

In this section we prove the convergence of the limit formulae in $C^m(\partial\Sigma)$-norm. In particular, this will be important in the last section for the proof of the main results of this dissertation. We need the following result, taken from [Gün57].

Lemma 4.3.1. Let Σ be an outer $C^{m+1,\alpha}$-domain and $F \in C^{n,\alpha}(\partial\Sigma)$, $0 \leq n \leq m$, $m \geq 0$, $0 < \beta < \alpha \leq 1$. Then we have

$$U_1[F] \in C^{n+1,\beta}(D), \tag{4.3}$$
$$U_1[F] \in C^{n+1,\beta}(\mathbb{R}^3\setminus\overline{D}). \tag{4.4}$$

Let Σ be an outer $C^{m+1,\alpha}$-domain and $F \in C^{n,\alpha}(\partial\Sigma)$, $0 \leq n \leq m+1$, $m \geq 0$, $0 < \beta < \alpha \leq 1$. Then we have

$$U_2[F] \in C^{n,\beta}(D), \tag{4.5}$$
$$U_2[F] \in C^{n,\beta}(\mathbb{R}^3\setminus\overline{D}). \tag{4.6}$$

Proof. This can be found in [Gün57, Paragraph II.19]. □

Remark 4.3.2. Note that even for integrable F we have that U_1 as well as U_2 are analytic outside of $\partial\Sigma$. But we need the Hölder continuity to assure the existence of an unique continuous continuation onto the boundary for U_1, U_2 and their derivatives.

Now we are able to prove the main result of this section.

Theorem 4.3.3. Let $m \in \mathbb{N}$, $m \geq 1$, $0 < \alpha \leq 1$ and Σ be an outer $C^{m+1,\alpha}$-domain. Furthermore let $F \in C^{m-1,\alpha}(\partial\Sigma)$ for $i = 1$, $F \in C^{m,\alpha}(\partial\Sigma)$ for $i = 2,3$ and $F \in C^{m+1,\alpha}(\partial\Sigma)$ for $i = 4$. Then we have $L_i^{\pm\tau}[F] \in C^m(\partial\Sigma)$ for $i = 1,2,3,4$ and all $\tau \in (0, \tau_0]$. Furthermore

$$\lim_{\tau \to 0^+} \|L_i^{\pm\tau}[F]\|_{C^m(\partial\Sigma)} = 0. \tag{4.7}$$

4.3. LIMIT FORMULAE IN $C^M(\partial\Sigma)$

Proof. Let $1 \leq m \in \mathbb{N}$ be arbitrary and the assumptions of the theorem be fulfilled. We define for $x \in \partial\Sigma$ and $\tau \in (0, \tau_0]$

$$V^1_{\pm\tau}[F](x) := \overline{U_1[F]}(x) - U_1[F](x \pm \tau\nu(x)),$$

$$V^2_{\pm\tau}[F](x) := \frac{\partial}{\partial\nu(x)}\overline{U_1[F]}(x) - \frac{\partial U_1}{\partial\nu(x)}[F](x \pm \tau\nu(x))$$

$$= \langle\nu(x), \nabla\overline{U_1[F]}(x)\rangle - \langle\nu(x \pm \tau\nu(x)), \nabla U_1[F](x \pm \tau\nu(x))\rangle$$

$$= \langle\nu(x), \nabla\overline{U_1[F]}(x)\rangle - \langle\nu(x), \nabla U_1[F](x \pm \tau\nu(x))\rangle$$

$$= \langle\nu(x), \nabla\overline{U_1[F]}(x) - \nabla U_1[F](x \pm \tau\nu(x))\rangle,$$

$$V^3_{\pm\tau}[F](x) := \overline{U_2[F]}(x) - U_2[F](x \pm \tau\nu(x)),$$

$$V^4_{\pm\tau}[F](x) := \frac{\partial}{\partial\nu(x)}\overline{U_2[F]}(x) - \frac{\partial U_2}{\partial\nu(x)}[F](x \pm \tau\nu(x))$$

$$= \langle\nu(x), \nabla\overline{U_2[F]}(x)\rangle - \langle\nu(x \pm \tau\nu(x)), \nabla U_2[F](x \pm \tau\nu(x))\rangle$$

$$= \langle\nu(x), \nabla\overline{U_2[F]}(x)\rangle - \langle\nu(x), \nabla U_2[F](x \pm \tau\nu(x))\rangle$$

$$= \langle\nu(x), \nabla\overline{U_2[F]}(x) - \nabla U_2[F](x \pm \tau\nu(x))\rangle,$$

where $\overline{U}$ denotes the continuation of the function U from Σ to $\Sigma \cup B_{\tau_0}(\partial\Sigma)$ for $+\tau$ or from D to $D \cup B_{\tau_0}(\partial\Sigma)$ for $-\tau$, respectively. This exists and is uniquely determined by Lemma 2.2.10 and Lemma 4.3.1. By the assumptions on F and Σ as well as Lemma 4.3.1, we have that $V^i_{\pm\tau}[F] \in C^m(\partial\Sigma)$ for all $\tau \in (0, \tau_0]$, $i = 1, 2, 3, 4$, because $V^i_{\pm\tau}[F](\Psi_j) \in C^m(B_1^{\mathbb{R}^2}(0))$ for $j = 1, \ldots, N$, as a composition of C^m-mappings. We will show that

$$\lim_{\tau \to 0+} \|V^i_{\pm\tau}[F]\|_{C^m(\partial\Sigma)} = 0,$$

for $i = 1, 2, 3, 4$. For $i = 1, 3$ it suffices to prove

$$\lim_{\tau \to 0+} \|W^{i,j,s}_{\pm\tau}[F]\|_{C^0(B_1^{\mathbb{R}^2}(0))} = 0,$$

for $i = 1, 3$, $j = 1, \ldots, N$ and $1 \leq |s| \leq m$, where

$$W^{1,j,s}_{\pm\tau}[F](x) := \partial_1^{s_1}\partial_2^{s_2}\Big(\overline{U_1[F]}(\Psi_j(x)) - U_1[F](\Psi_j(x) \pm \tau\nu(\Psi_j(x)))\Big),$$

$$W^{3,j,s}_{\pm\tau}[F](x) := \partial_1^{s_1}\partial_2^{s_2}\Big(\overline{U_2[F]}(\Psi_j(x)) - U_2[F](\Psi_j(x) \pm \tau\nu(\Psi_j(x)))\Big),$$

for all $x \in B_1^{\mathbb{R}^2}(0)$, $\tau \in (0, \tau_0]$, $1 \leq s_1 + s_2 \leq m$ and $j = 1, \ldots, N$. For simplicity we denoted $\Psi_j(x, 0)$ by $\Psi_j(x)$. For $i = 2, 4$ we have

$$\|V^2_{\pm\tau}[F]\|_{C^m(\partial\Sigma)} = \|\langle\nu(\,\cdot\,), \nabla\overline{U_1[F]}(\,\cdot\,) - \nabla U_1[F](\,\cdot\, \pm \tau\nu(\,\cdot\,))\rangle\|_{C^m(\partial\Sigma)}$$

$$\leq \|\nu\|_{C^m(\partial\Sigma)} \cdot \sum_{k=1}^{3} \|\partial_k \overline{U_1[F]}(\,\cdot\,) - \partial_k U_1[F](\,\cdot\, \pm \tau\nu(\,\cdot\,))\|_{C^m(\partial\Sigma)},$$

$$\|V_{\pm\tau}^4[F]\|_{C^m(\partial\Sigma)} = \|\langle \nu(\,\cdot\,), \nabla\overline{U_2[F]}(\,\cdot\,) - \nabla U_2[F](\,\cdot\, \pm \tau\nu(\,\cdot\,))\rangle\|_{C^m(\partial\Sigma)}$$

$$\leq \|\nu\|_{C^m(\partial\Sigma)} \cdot \sum_{k=1}^{3} \|\partial_k \overline{U_2[F]}(\,\cdot\,) - \partial_k U_2[F](\,\cdot\, \pm \tau\nu(\,\cdot\,))\|_{C^m(\partial\Sigma)},$$

Therefore it suffices in these two cases to prove that

$$\lim_{\tau \to 0^+} \|W_{\pm\tau}^{i,j,s}[F]\|_{C^0(B_1^{\mathbb{R}^2}(0))} = 0,$$

for $i = 2, 4$, $j = 1, \ldots, N$ and $1 \leq |s| \leq m$, where

$$W_{\pm\tau}^{2,j,s}[F](x) := \partial_1^{s_1} \partial_2^{s_2}\left(\partial_k \overline{U_1[F]}(\Psi_j(x)) - \partial_k U_1[F](\Psi_j(x) \pm \tau\nu(\Psi_j(x)))\right),$$

$$W_{\pm\tau}^{4,j,s}[F](x) := \partial_1^{s_1} \partial_2^{s_2}\left(\partial_k \overline{U_2[F]}(\Psi_j(x)) - \partial_k U_2[F](\Psi_j(x) \pm \tau\nu(\Psi_j(x)))\right),$$

for all $x \in B_1^{\mathbb{R}^2}(0)$, $\tau \in (0, \tau_0]$, $1 \leq s_1 + s_2 \leq m$, $j = 1, \ldots, N$ and $k = 1, 2, 3$. We can treat all cases, independent of i, k and s, at the same time if we set

$$Z_{\pm\tau}^{j,s}[F] := \partial_1^{s_1} \partial_2^{s_2}\left(\overline{S_i[F]}(\Psi_j) - US_i[F](\Psi_j \pm \tau\nu(\Psi_j))\right),$$

for all $\tau \in (0, \tau_0]$, where S_i can be obtained from the definitions of $W_{\pm\tau}^{i,j,s}$ above. In any case we find $\overline{S_i}(\Psi_j), S_i(\Psi_j \pm \tau\nu(\Psi_j)) \in C^m(B_1^{\mathbb{R}^2}(0))$, by the assumptions on F and Σ due to Lemma 4.3.1 in combination with Lemma 2.2.10. We have

$$\|Z_{\pm\tau}^{j,s}[F]\|_{C^0(B_1^{\mathbb{R}^2}(0))}$$
$$= \|\partial_1^{s_1} \partial_2^{s_2}\left(\overline{S_i[F]}(\Psi_j) - S_i[F](\Psi_j \pm \tau\nu(\Psi_j))\right)\|_{C^0(B_1^{\mathbb{R}^2}(0))}$$
$$\leq C_1 \Bigg(\sum_{0 \leq |r| \leq |s|, |t^1| + \ldots + |t^{|r|}| = |s|} \|\left(\partial_1^{r_1} \partial_2^{r_2} \partial_3^{r_3} \overline{S_i[F]}\right)(\Psi_j) \cdot \partial_1^{t_1^1} \partial_2^{t_2^1} \Psi_i \cdot \ldots \cdot \partial_1^{t_1^{|r|}} \partial_2^{t_2^{|r|}} \Psi_j$$
$$- \left(\partial_1^{r_1} \partial_2^{r_2} \partial_3^{r_3} S_i[F]\right)(\Psi_j \pm \tau\nu(\Psi_j)) \cdot \partial_1^{t_1^1} \partial_2^{t_2^1}(\Psi_j \pm \tau\nu(\Psi_j)) \cdot \ldots \cdot \partial_1^{t_1^{|r|}} \partial_2^{t_2^{|r|}}(\Psi_j \pm \tau\nu(\Psi_j))\|_{C^0(B_1^{\mathbb{R}^2}(0))}$$
$$\leq C_1 \Bigg(\sum_{0 \leq |r| \leq |s|, |t^1| + \ldots + |t^{|r|}| = |s|} \|\left(\partial_1^{r_1} \partial_2^{r_2} \partial_3^{r_3} \overline{S_i[F]}\right)(\Psi_j) \cdot \partial_1^{t_1^1} \partial_2^{t_2^1} \Psi_i \cdot \ldots \cdot \partial_1^{t_1^{|r|}} \partial_2^{t_2^{|r|}} \Psi_j$$
$$- \left(\partial_1^{r_1} \partial_2^{r_2} \partial_3^{r_3} S_i[F]\right)(\Psi_j \pm \tau\nu(\Psi_j)) \cdot \partial_1^{t_1^1} \partial_2^{t_2^1} \Psi_j \cdot \ldots \cdot \partial_1^{t_1^{|r|}} \partial_2^{t_2^{|r|}} \Psi_j\|_{C^0(B_1^{\mathbb{R}^2}(0))}$$
$$+ \sum_{0 \leq |r| \leq |s|, |t^1| + \ldots + |t^{|r|}| = |s|} \sum_{1 \leq p \leq |r|} \tau^p \|\left(\partial_1^{r_1} \partial_2^{r_2} \partial_3^{r_3} S_i[F]\right)(\Psi_j \pm \tau\nu(\Psi_j)) \cdot$$

4.3. LIMIT FORMULAE IN $C^M(\partial\Sigma)$

$$\partial_1^{t_1^1}\partial_2^{t_2^1}\big(\nu(\Psi_j)\big)\cdot\ldots\cdot\partial_1^{t_1^p}\partial_2^{t_2^p}\big(\nu(\Psi_j)\big)\cdot\partial_1^{t_1^{p+1}}\partial_2^{t_2^{p+1}}\Psi_j\cdot\ldots\cdot\partial_1^{t_1^{|r|}}\partial_2^{t_2^{|r|}}\Psi_j\|_{C^0(B_1^{\mathbb{R}^2}(0))}\Big)$$

$$\leq C_1\Bigg(\binom{|s|}{|r|}\sum_{0\leq|r|\leq|s|}\sup_{x\in\partial\Sigma}\Big(\partial_1^{r_1}\partial_2^{r_2}\partial_3^{r_3}\overline{S_i[F]}(x)-\big(\partial_1^{r_1}\partial_2^{r_2}\partial_3^{r_3}S_i[F]\big)(x\pm\tau\nu(x))\Big)$$

$$\cdot\big(\|\Psi_j\|_{C^m(B_1^{\mathbb{R}^2}(0))}+1\big)^m$$

$$+\sum_{0\leq|r|\leq|s|,|t^1|+\ldots+|t^{|r|}|=|s|}\sum_{1\leq p\leq|r|}\tau^p\|\big(\partial_1^{r_1}\partial_2^{r_2}\partial_3^{r_3}S_i[F]\big)(\Psi_j\pm\tau\nu(\Psi_j))\cdot$$

$$\partial_1^{t_1^1}\partial_2^{t_2^1}\big(\nu(\Psi_j)\big)\cdot\ldots\cdot\partial_1^{t_1^p}\partial_2^{t_2^p}\big(\nu(\Psi_j)\big)\cdot\partial_1^{t_1^{p+1}}\partial_2^{t_2^{p+1}}\Psi_j\cdot\ldots\cdot\partial_1^{t_1^{|r|}}\partial_2^{t_2^{|r|}}\Psi_j\|_{C^0(B_1^{\mathbb{R}^2}(0))}\Bigg),$$

where we used the triangle inequality and broke $\partial_1^{s_1}\partial_2^{s_2}\big(\overline{U_i[F]}(\Psi_j)-U_i[F](\Psi_j\pm\tau\nu(\Psi_j))\big)$ down to a sum of terms of the form

$$\big(\partial_1^{r_1}\partial_2^{r_2}\partial_3^{r_3}\overline{U_i[F]}\big)(\Psi_j)\cdot\partial_1^{t_1^1}\partial_2^{t_2^1}\Psi_j\cdot\ldots\cdot\partial_1^{t_1^{|r|}}\partial_2^{t_2^{|r|}}\Psi_j, \tag{4.8}$$

$$\big(\partial_1^{r_1}\partial_2^{r_2}\partial_3^{r_3}U_i[F]\big)(\Psi_j\pm\tau\nu(\Psi_j))\cdot\partial_1^{t_1^1}\partial_2^{t_2^1}\Psi_j\cdot\ldots\cdot\partial_1^{t_1^l}\partial_2^{t_2^l}\Psi_j, \tag{4.9}$$

$$\big(\partial_1^{r_1}\partial_2^{r_2}\partial_3^{r_3}U_i[F]\big)(\Psi_j\pm\tau\nu(\Psi_j))\cdot\partial_1^{t_1^1}\partial_2^{t_2^1}\big(\nu(\Psi_j)\big)\cdot\ldots\cdot\partial_1^{t_1^p}\partial_2^{t_2^p}\big(\nu(\Psi_j)\big)$$
$$\cdot\partial_1^{t_1^{p+1}}\partial_2^{t_2^{p+1}}\Psi_j\cdot\ldots\cdot\partial_1^{t_1^{|r|}}\partial_2^{t_2^{|r|}}\Psi_j, \tag{4.10}$$

using the chain and product rule of differentiation. C_1 is the maximal multiplicity which a single term of (4.8), (4.9) or (4.10) can posses. Furthermore we added 1 for the case that $\|\Psi_j\|_{C^m(B_1^{\mathbb{R}^2}(0))}<1$. Note that we have $\|\Psi_j\|_{C^m(B_1^{\mathbb{R}^2}(0))}<\infty$ in any case, because $\Psi_j\in C^m\big(\overline{B_1^{\mathbb{R}^2}(0)}\big)$ for all $j\in\{1,\ldots,N\}$. Now all terms in the first sum converge to zero if τ does, because $\partial_1^{r_1}\partial_2^{r_2}\partial_3^{r_3}U_i[F]$ is, as a continuous function, uniformly continuous on the compact sets $\overline{D}$ and $\overline{\Sigma\cap B_R^{\mathbb{R}^3}(0)}$, where $R>0$ is such large that $B_{\tau_0}(\partial\Sigma)\subset B_R^{\mathbb{R}^3}(0)$. Each single norm in the second sum can be estimated by

$$\|S_i[F]\|_{C^m(\overline{D})}\cdot\big(\|\nu\|_{C^m(\partial\Sigma)}+1\big)^m\cdot\big(\|\Psi_j\|_{C^m(B_1^{\mathbb{R}^2}(0))}+1\big)^m, \text{ for }-\tau,$$

$$\|S_i[F]\|_{C^m\left(\overline{\Sigma\cap B_R^{\mathbb{R}^3}(0)}\right)}\cdot\big(\|\nu\|_{C^m(\partial\Sigma)}+1\big)^m\cdot\big(\|\Psi_j\|_{C^m(B_1^{\mathbb{R}^2}(0))}+1\big)^m, \text{ for }+\tau.$$

Thus the second sum also converges to zero as τ does. Consequently we proved

$$\lim_{\tau\to 0^+}\|Z_{\pm\tau}^{s,j}[F]\|_{C^0(B_1^{\mathbb{R}^2}(0))}=0,$$

for all $1\leq|s|\leq m$ $j=1,\ldots,N$ and $k=1,2,3$. Using Lemma 4.2.2 for $m=0$ we have proved

$$\lim_{\tau\to 0^+}\|V_{\pm\tau}^i[F]\|_{C^m(\partial\Sigma)}=0.$$

Finally Lemma 2.4.6, or Lemma 2.4.7 respectively, gives

$$\overline{U_1[F]}(x) = U_1[F](x),$$
$$\frac{\partial}{\partial \nu}\overline{U_1[F]}(x) = \frac{\partial U_1}{\partial \nu}[F](x) \mp 2\pi F(x),$$
$$\overline{U_2[F]}(x) = U_2[F](x) \pm 2\pi F(x),$$
$$\frac{\partial}{\partial \nu}\overline{U_2[F]}(x) = \frac{\partial U_2}{\partial \nu}[F](x),$$

for all $x \in \partial\Sigma$, which yields

$$\lim_{\tau \to 0^+} \|L_i^{\pm\tau}[F]\|_{C^m(\partial\Sigma)} = 0.$$

So we assume the theorem to be proved. □

4.4 Limit Formulae in Hölder Norms

In this section we prove that, under slightly stronger assumption on Σ, the convergence of the limit formulae in $C^m(\partial\Sigma)$ even holds in $C^{m,\beta}(\partial\Sigma)$-norm. In order to prove it, we again use Lemma 4.3.1. We come directly to the main result of this section.

Theorem 4.4.1. *Let* $m \in \mathbb{N}$, $m \geq 0$, $0 < \beta < \alpha \leq 1$, Σ *be an outer* C^{m+2}-*domain. Furthermore, let* $F \in C^{m-1,\alpha}(\partial\Sigma)$ *for* $i = 1$, $m \geq 1$, $F \in C^{0,\alpha}(\partial\Sigma)$ *for* $i = 1$, $m = 0$, $F \in C^{m,\alpha}(\partial\Sigma)$ *for* $i = 2, 3$ *and* $F \in C^{m+1,\alpha}(\partial\Sigma)$ *for* $i = 4$. *Then we have* $L_i^{\pm\tau}[F] \in C^{m,\beta}(\partial\Sigma)$ *for* $i = 1, 2, 3, 4$ *and all* $\tau \in (0, \tau_0]$. *Furthermore*

$$\lim_{\tau \to 0^+} \|L_i^{\pm\tau}[F]\|_{C^{m,\beta}(\partial\Sigma)} = 0. \tag{4.11}$$

Proof. At first we introduce the following equivalent definition of the Hölder constant by

$$\text{höl}_\beta^*(f) := \sup\left\{\frac{|f(y_1) - f(y_2)|}{|y_1 - y_2|^\beta}\bigg|\, y_1, y_2 \in \partial\Sigma, y_1 \neq y_2\right\}$$

for all $f \in C^{0,\beta}(\partial\Sigma)$. In order to show the equivalence, we have to find constants $0 < C_2^1, C_2^2 < \infty$ such that $\text{höl}_\beta(f) \leq C_2^1 \text{höl}_\beta^*(f)$ and $\text{höl}_\beta^*(f) \leq C_2^2 \text{höl}_\beta(f)$ for all $f \in C^{0,\beta}(\partial\Sigma)$, where $\text{höl}_\beta(f)$ is the Hölder constant of $C^{0,\beta}(\partial\Sigma)$, introduced by Definition 2.2.8. Because $\partial\Sigma$ is an outer

4.4. LIMIT FORMULAE IN HÖLDER NORMS

C^2-domain, we have that Ψ_i and Ψ_i^{-1} are $C^{0,1}$-functions, for $i = 1, \ldots, N$, and consequently we find constants $0 < C_3^i, C_4^i < \infty$ with

$$|\Psi_i(x_1, 0) - \Psi_i(x_2, 0)| \leq C_3^i |(x_1, 0) - (x_2, 0)|,$$
$$|\Psi_i^{-1}(y_1) - \Psi_i^{-1}(y_2)| \leq C_4^i |y_1 - y_2|,$$

for all $x_1, x_2 \in B_1^{\mathbb{R}^2}(0)$ and $y_1, y_2 \in U_i$. Now we estimate

$$\text{höl}_\beta(f) = \sup\left\{ \frac{|f(\Psi_i(x_1, 0)) - f(\Psi_i(x_2, 0))|}{|(x_1, 0) - (x_2, 0)|^\beta} \Big| x_1, x_2 \in B_1^{\mathbb{R}^2}(0), x_1 \neq x_2 \right\}$$

$$\leq \sup\left\{ \frac{|f(\Psi_i(x_1, 0)) - f(\Psi_i(x_2, 0))|}{\left(\frac{1}{C_3^i}\right)^\beta |\Psi_i(x_1, 0) - \Psi_i(x_2, 0)|^\beta} \Big| x_1, x_2 \in B_1^{\mathbb{R}^2}(0), x_1 \neq x_2 \right\}$$

$$= \left(C_3^i\right)^\beta \sup\left\{ \frac{|f(y_1) - f(y_2)|}{|y_1 - y_2|^\beta} \Big| y_1, y_2 \in \partial\Sigma, y_1 \neq y_2 \right\} = \text{höl}_\beta^*(f),$$

and consequently for $C_2^1 := 1 + \sum_{i=1}^N (C_3^i)^\beta$ the desired condition is fulfilled. In turn we have

$$\text{höl}_\beta^*(f) = \sup\left\{ \frac{|f(y_1) - f(y_2)|}{|y_1 - y_2|^\beta} \Big| y_1, y_2 \in \partial\Sigma, y_1 \neq y_2 \right\}$$

$$\leq \sum_{i=1}^N \sup\left\{ \frac{|f(\Psi_i(x_1, 0)) - f(\Psi_i(x_2, 0))|}{|\Psi_i(x_1, 0) - \Psi_i(x_2, 0)|^\beta} \Big| x_1, x_2 \in B_1^{\mathbb{R}^2}(0), x_1 \neq x_2 \right\}$$

$$\leq \sup\left\{ \frac{|f(\Psi_i(x_1, 0)) - f(\Psi_i(x_2, 0))|}{\left(\frac{1}{C_4^i}\right)^\beta |x_1 - x_2|^\beta} \Big| x_1, x_2 \in B_1^{\mathbb{R}^2}(0), x_1 \neq x_2 \right\} = \text{höl}_\beta(f),$$

and $C_2^2 := 1 + \sum_{i=1}^N (C_4^i)^\beta$ is a possible choice. In this proof we will work with this equivalent definition of the Hölder constant, neglecting the equivalent constants to simplify the exposition. Now we come to the proof of the theorem. Due to Lemma 4.3.1 together with the assumptions on F and $\partial\Sigma$, we have $U_1[F], U_2[F] \in C^{m,\beta}(D)$, $U_1[F], U_2[F] \in C^{m,\beta}(\Sigma)$, for $i = 1, 3$, and $\nabla U_1[F], \nabla U_2[F] \in C^{m,\beta}(D)$, $\nabla U_1[F], \nabla U_2[F] \in C^{m,\beta}(\Sigma)$, for $i = 2, 4$. Note the $\nabla U \in C^m(\Sigma)$ means $\partial_k U \in C^m(\Sigma)$ for $k = 1, 2, 3$. Furthermore, Lemma 2.2.10 gives continuations $U_1[F], U_2[F] \in C^{m,\beta}(D \cup B_{\tau_0}(\partial\Sigma))$, $U_1[F], U_2[F] \in C^{m,\beta}(\Sigma \cup B_{\tau_0}(\partial\Sigma))$, for $i = 1, 3$, and $\nabla U_1[F], \nabla U_2[F] \in C^{m,\beta}(D \cup B_{\tau_0}(\partial\Sigma))$, $\nabla U_1[F], \nabla U_2[F] \in C^{m,\beta}(\Sigma \cup B_{\tau_0}(\partial\Sigma))$, for $i = 2, 4$. Moreover, we have

$$\text{höl}_\beta(U_i[F](\psi_j \pm \tau\nu(\Psi_j)))$$

$$\leq c_4^1 C_1 \left(\sum_{0 \leq |r| \leq |s|, |t^1| + \ldots + |t^{|r|}| = |s|} \mathrm{höl}_\beta \left(\left(\partial_1^{r_1} \partial_2^{r_2} \partial_3^{r_3} U_i[F] \right) (\Psi_j \pm \tau \nu(\Psi_j)) \cdot \partial_1^{t_1^1} \partial_2^{t_2^1} \Psi_j \cdot \ldots \cdot \partial_1^{t_1^{|r|}} \partial_2^{t_2^{|r|}} \Psi_j \right) \right.$$

$$+ \sum_{0 \leq |r| \leq |s|, |t^1| + \ldots + |t^{|r|}| = |s|} \sum_{1 \leq p \leq |r|} \tau^p \cdot \mathrm{höl}_\beta \left(\left(\partial_1^{r_1} \partial_2^{r_2} \partial_3^{r_3} U_i[F] \right) (\Psi_j \pm \tau \nu(\Psi_j)) \cdot \right.$$

$$\left. \left. \partial_1^{t_1^1} \partial_2^{t_2^1} \big(\nu(\Psi_j) \big) \cdot \ldots \cdot \partial_1^{t_1^p} \partial_2^{t_2^p} \big(\nu(\Psi_j) \big) \cdot \partial_1^{t_1^{p+1}} \partial_2^{t_2^{p+1}} \Psi_j \cdot \ldots \cdot \partial_1^{t_1^{|r|}} \partial_2^{t_2^{|r|}} \Psi_j \right) \right)$$

$$\leq c_4^1 c_4^2 C_1 \left(\left(\|\Psi_j\|_{C^{m,\beta}(B_1^{\mathbb{R}^2}(0))} + 1 \right)^m \cdot \sum_{0 \leq |r| \leq |s|} \binom{m}{|r|} \cdot \mathrm{höl}_\beta \left(\left(\partial_1^{r_1} \partial_2^{r_2} \partial_3^{r_3} U_i[F] \right) (\Psi_j \pm \tau \nu(\Psi_j)) \right) \right.$$

$$+ \left(\|\Psi_j\|_{C^{m,\beta}(B_1^{\mathbb{R}^2}(0))} + 1 \right)^m \cdot \left(\|\nu\|_{C^{m,\beta}(\partial \Sigma)} + 1 \right)^m$$

$$\left. \sum_{0 \leq |r| \leq |s|} \binom{m}{|r|} \sum_{1 \leq p \leq |r|} \tau^p \cdot \mathrm{höl}_\beta \left(\left(\partial_1^{r_1} \partial_2^{r_2} \partial_3^{r_3} U_i[F] \right) (\Psi_j \pm \tau \nu(\Psi_j)) \right) \right)$$

$$\leq c_4^1 c_4^2 C_1 \left(\|\Psi_j\|_{C^{m,\beta}(B_1^{\mathbb{R}^2}(0))} + 1 \right)^m \cdot \left(\|\nu\|_{C^{m,\beta}(\partial \Sigma)} + 1 \right)^m$$

$$\cdot \left(\sum_{0 \leq |r| \leq |s|} \binom{m}{|r|} \cdot \mathrm{höl}_\beta \left(\left(\partial_1^{r_1} \partial_2^{r_2} \partial_3^{r_3} U_i[F] \right) (\Psi_j \pm \tau \nu(\Psi_j)) \right) \right.$$

$$\left. + \sum_{0 \leq |r| \leq |s|} \binom{m}{|r|} \sum_{1 \leq p \leq |r|} \tau^p \cdot \mathrm{höl}_\beta \left(\left(\partial_1^{r_1} \partial_2^{r_2} \partial_3^{r_3} U_i[F] \right) (\Psi_j \pm \tau \nu(\Psi_j)) \right) \right)$$

$$= c_4^1 c_4^2 C_1 \left(\|\Psi_j\|_{C^{m,\beta}(B_1^{\mathbb{R}^2}(0))} + 1 \right)^m \cdot \left(\|\nu\|_{C^{m,\beta}(\partial \Sigma)} + 1 \right)^m$$

$$\cdot \left(\sum_{0 \leq |r| \leq |s|} \binom{m}{|r|} \cdot \mathrm{höl}_\beta^* \left(\left(\partial_1^{r_1} \partial_2^{r_2} \partial_3^{r_3} U_i[F] \right) (\,\cdot\, \pm \tau \nu(\,\cdot\,)) \right) \right.$$

$$\left. + \sum_{0 \leq |r| \leq |s|} \binom{m}{|r|} \sum_{1 \leq p \leq |r|} \tau^p \cdot \mathrm{höl}_\beta^* \left(\left(\partial_1^{r_1} \partial_2^{r_2} \partial_3^{r_3} U_i[F] \right) (\,\cdot\, \pm \tau \nu(\,\cdot\,)) \right) \right),$$

where c_4^i are the constants from Lemma 2.2.9 and C_1, U_i are taken from the proof of Theorem 4.3.3. This estimate holds also for $\tau = 0$, if we replace U_i by $\overline{U_i}$. In any case, the Hölder constants can be estimated by $c_4^3 c_4^4 \|U_i\|_{C^{m,\beta}(\overline{\Sigma \cap B_{\tau_0}(\partial \Sigma)})}$, or $c_4^3 c_4^4 \|U_i\|_{C^{m,\beta}(\overline{D})}$ respectively. Thus we find $U_1[F](\,\cdot\, \pm \tau \nu(\,\cdot\,)), U_2[F](\,\cdot\, \pm \tau \nu(\,\cdot\,)) \in C^{m,\beta}(\partial \Sigma)$ for all $\tau \in [0, \tau_0]$. With the same reasoning, we additionally obtain that $\partial_k U_1[F](\,\cdot\, \pm \tau \nu(\,\cdot\,)), \partial_k U_2[F](\,\cdot\, \pm \tau \nu(\,\cdot\,)) \in C^{m,\beta}(\partial \Sigma)$, $k = 1, 2, 3$, $\tau \in [0, \tau_0]$ and thus we have that the functions $\langle \nu(\,\cdot\,), \nabla U_1[F](\,\cdot\, \pm \tau \nu(\,\cdot\,)) \rangle$, $\langle \nu(\,\cdot\,), \nabla U_2[F](\,\cdot\, \pm \tau \nu(\,\cdot\,)) \rangle$ are elements of $C^{m,\beta}(\partial \Sigma)$ for all $\tau \in [0, \tau_0]$, because of Lemma 2.2.9 and the fact that $\nu \in C^{m,\beta}(\partial \Sigma)$ for an outer C^{m+2}-domain, see

4.4. LIMIT FORMULAE IN HÖLDER NORMS

Lemma 2.1.6 and Lemma 2.2.9. For $i = 2, 4$ we have

$$\|\langle \nu(\,\cdot\,), \nabla \overline{U_1[F]}(\,\cdot\,) - \nabla U_1[F](\,\cdot \pm \tau\nu(\,\cdot\,))\rangle\|_{C^{m,\beta}(\partial\Sigma)}$$

$$\leq c_4^1 \cdot c_4^2 \cdot \|\nu\|_{C^{m,\beta}(\partial\Sigma)} \cdot \sum_{j=1}^{3} \|\partial_j \overline{U_1[F]}(\,\cdot\,) - \partial_j U_1[F](\,\cdot \pm \tau\nu(\,\cdot\,))\|_{C^{m,\beta}(\partial\Sigma)}$$

$$\|\langle \nu(\,\cdot\,), \nabla \overline{U_2[F]}(\,\cdot\,) - \nabla U_2[F](\,\cdot \pm \tau\nu(\,\cdot\,))\rangle\|_{C^{m,\beta}(\partial\Sigma)}$$

$$\leq c_4^1 \cdot c_4^2 \cdot \|\nu\|_{C^{m,\beta}(\partial\Sigma)} \cdot \sum_{j=1}^{3} \|\partial_j \overline{U_2[F]}(\,\cdot\,) - \partial_j U_2[F](\,\cdot \pm \tau\nu(\,\cdot\,))\|_{C^{m,\beta}(\partial\Sigma)},$$

with the constants from Lemma 2.2.9. So taking into account the results from Theorem 4.3.3, it is left prove that

$$\lim_{\tau \to 0} \text{höl}_\beta(W_{\pm\tau}^{i,j,s}[F]) = 0.$$

for all $|s| = m$, $i = 1, 2, 3, 4$ and $j = 1, \ldots, N$, where $W_{\pm\tau}^{i,j,s}[F]$ is taken from the proof of Theorem 4.3.3. Then

$$\lim_{\tau \to 0^+} \|V_{\pm\tau}^i[F]\|_{C^{m,\beta}(\partial\Sigma)} = 0,$$

and with the same reasoning as in the previous proof we get

$$\lim_{\tau \to 0^+} \|L_i^{\pm\tau}[F]\|_{C^{m,\beta}(\partial\Sigma)} = 0.$$

Using the definition of $Z_{\pm\tau}^{j,s}[F]$, we estimate similar as above to get

$$\text{höl}_\beta(Z_{\pm\tau}^{j,s}[F])$$

$$\leq c_4^1 C_1 \Bigg(\sum_{0 \leq |r| \leq |s|, |t^1| + \ldots + |t^{|r|}| = |s|} \text{höl}_\beta \bigg(\Big(\partial_1^{r_1} \partial_2^{r_2} \partial_3^{r_3} \overline{S_i[F]}\Big)(\Psi_j) \cdot \partial_1^{t_1^1} \partial_2^{t_2^1} \Psi_i \cdot \ldots \cdot \partial_1^{t_1^{|r|}} \partial_2^{t_2^{|r|}} \Psi_j$$

$$- \Big(\partial_1^{r_1} \partial_2^{r_2} \partial_3^{r_3} S_i[F]\Big)(\Psi_j \pm \tau\nu(\Psi_j)) \cdot \partial_1^{t_1^1} \partial_2^{t_2^1} \Psi_j \cdot \ldots \cdot \partial_1^{t_1^{|r|}} \partial_2^{t_2^{|r|}} \Psi_j \bigg)$$

$$+ \sum_{0 \leq |r| \leq |s|, |t^1| + \ldots + |t^{|r|}| = |s|} \sum_{1 \leq p \leq |r|} \tau^p \cdot \text{höl}_\beta \bigg(\Big(\partial_1^{r_1} \partial_2^{r_2} \partial_3^{r_3} S_i[F]\Big)(\Psi_j \pm \tau\nu(\Psi_j)) \cdot$$

$$\partial_1^{t_1^1} \partial_2^{t_2^1} \Big(\nu(\Psi_j)\Big) \cdot \ldots \cdot \partial_1^{t_1^p} \partial_2^{t_2^p} \Big(\nu(\Psi_j)\Big) \cdot \partial_1^{t_1^{p+1}} \partial_2^{t_2^{p+1}} \Psi_j \cdot \ldots \cdot \partial_1^{t_1^{|r|}} \partial_2^{t_2^{|r|}} \Psi_j \bigg) \Bigg)$$

$$\leq c_4^1 c_4^2 C_1 \Big(\big(\|\Psi_j\|_{C^{m,\beta}(B_1^{\mathbb{R}^2}(0))}\big) + 1 \Big)^m$$

$$\sum_{0\leq |r|\leq |s|}\binom{m}{|r|}\cdot \mathrm{h\ddot{o}l}_\beta\bigg(\Big(\partial_1^{r_1}\partial_2^{r_2}\partial_3^{r_3}\overline{S_i[F]}\Big)(\Psi_j) - \Big(\partial_1^{r_1}\partial_2^{r_2}\partial_3^{r_3}S_i[F]\Big)(\Psi_j \pm \tau\nu(\Psi_j))\bigg)$$

$$+ \big(\|\Psi_j\|_{C^{m,\beta}(B_1^{\mathbb{R}^2}(0))} + 1\big)^m \cdot \big(\|\nu\|_{C^{m,\beta}(\partial\Sigma)} + 1\big)^m$$

$$\sum_{0\leq |r|\leq |s|}\binom{m}{|r|}\sum_{1\leq p\leq |r|}\tau^p\cdot \mathrm{h\ddot{o}l}_\beta\bigg(\Big(\partial_1^{r_1}\partial_2^{r_2}\partial_3^{r_3}S_i[F]\Big)(\Psi_j \pm \tau\nu(\Psi_j))\bigg)$$

$$\leq c_4^1 c_4^2 C_1 \big(\|\Psi_j\|_{C^{m,\beta}(B_1^{\mathbb{R}^2}(0))} + 1\big)^m \cdot \big(\|\nu\|_{C^{m,\beta}(\partial\Sigma)} + 1\big)^m$$

$$\cdot \Bigg(\sum_{0\leq |r|\leq |s|}\binom{m}{|r|}\cdot \mathrm{h\ddot{o}l}_\beta\bigg(\Big(\partial_1^{r_1}\partial_2^{r_2}\partial_3^{r_3}\overline{S_i[F]}\Big)(\Psi_j) - \Big(\partial_1^{r_1}\partial_2^{r_2}\partial_3^{r_3}S_i[F]\Big)(\Psi_j \pm \tau\nu(\Psi_j))\bigg)$$

$$+ \sum_{0\leq |r|\leq |s|}\binom{m}{|r|}\sum_{1\leq p\leq |r|}\tau^p\cdot \mathrm{h\ddot{o}l}_\beta\bigg(\Big(\partial_1^{r_1}\partial_2^{r_2}\partial_3^{r_3}S_i[F]\Big)(\Psi_j \pm \tau\nu(\Psi_j))\bigg)\Bigg)$$

$$= c_4^1 c_4^2 C_1 \big(\|\Psi_j\|_{C^{m,\beta}(B_1^{\mathbb{R}^2}(0))} + 1\big)^m \cdot \big(\|\nu\|_{C^{m,\beta}(\partial\Sigma)} + 1\big)^m$$

$$\cdot \Bigg(\sum_{0\leq |r|\leq |s|}\binom{m}{|r|}\cdot \mathrm{h\ddot{o}l}_\beta^*\bigg(\Big(\partial_1^{r_1}\partial_2^{r_2}\partial_3^{r_3}\overline{S_i[F]}\Big)(\,\cdot\,) - \Big(\partial_1^{r_1}\partial_2^{r_2}\partial_3^{r_3}S_i[F]\Big)(\,\cdot\, \pm \tau\nu(\,\cdot\,))\bigg)$$

$$+ \sum_{0\leq |r|\leq |s|}\binom{m}{|r|}\sum_{1\leq p\leq |r|}\tau^p\cdot \mathrm{h\ddot{o}l}_\beta^*\bigg(\Big(\partial_1^{r_1}\partial_2^{r_2}\partial_3^{r_3}S_i[F]\Big)(\,\cdot\, \pm \tau\nu(\,\cdot\,))\bigg)\Bigg),$$

where c_4^i are again the constants from Lemma 2.2.9 and as well C_1 as S_i are taken from the proof of Theorem 4.3.3. The Hölder constants in the second sum can be estimated by $c_4^3\|S_i\|_{C^{m,\beta}(\overline{\Sigma\cap B_{\tau_0}(\partial\Sigma)})}$, or $c_4^3\|S_i\|_{C^{m,\beta}(\overline{D})}$ respectively, and consequently the sum converges to zero as τ does. So it is left to show

$$\lim_{\tau\to 0}\mathrm{h\ddot{o}l}_\beta^*\Big(\partial_1^{r_1}\partial_2^{r_2}\partial_3^{r_3}\overline{S_i[F]}(\,\cdot\,) - \partial_1^{r_1}\partial_2^{r_2}\partial_3^{r_3}S_i[F](\,\cdot\, \pm \tau\nu(\,\cdot\,))\Big) = 0.$$

for all $0 \leq |r| \leq m$. In each case we have that $\partial_1^{r_1}\partial_2^{r_2}\partial_3^{r_3}S_i[F] \in C^{0,\beta}(\overline{D})$ and $\partial_1^{r_1}\partial_2^{r_2}\partial_3^{r_3}S_i[F] \in C^{0,\beta}(\overline{\Sigma\cap B_{\tau_0}(\partial\Sigma)})$, for all $0 < \beta < \alpha$. So it suffices to show that for each $\tilde{U}_{\pm\tau} := \tilde{U}(x) - \tilde{U}(x\pm\tau\nu(x))$, $x \in \partial\Sigma$, with $\tilde{U} \in C^{0,\beta}(\overline{D})$ and $\tilde{U} \in C^{0,\beta}(\overline{\Sigma\cap B_{\tau_0}(\partial\Sigma)})$ for all $0 < \beta < \alpha$, we find

$$\lim_{\tau\to 0}\mathrm{h\ddot{o}l}_\beta^*(\tilde{U}_{\pm\tau}) = \lim_{\tau\to 0}\left(\sup\left\{\frac{|\tilde{U}_{\pm\tau}(x) - \tilde{U}_{\pm\tau}(y)|}{|x-y|^\beta}\bigg| x,y\in\partial\Sigma, x\neq y\right\}\right) = 0,$$

and we are done with this proof. We have that

$$\mathrm{h\ddot{o}l}_\beta^*(\tilde{U}_{\pm\tau}) \leq 2\cdot\max\left\{\mathrm{h\ddot{o}l}_{\overline{D}}^\beta(\tilde{U}), \mathrm{h\ddot{o}l}_{\overline{\Sigma}}^\beta(\tilde{U})\right\} < \infty,$$

4.4. LIMIT FORMULAE IN HÖLDER NORMS

for all $\tau \in [0, \tau_0]$, so we have at least the existence of the lim sup. Now assume

$$\lim\sup\nolimits_{\tau \to 0} \text{höl}_\beta^*(\tilde{U}_{\pm\tau}) = C_5 > 0.$$

We can choose a sequence $(\tau_n)_{n \in \mathbb{N}}$ with $\tau_n \to 0$ for $n \to \infty$ and

$$\lim_{n \to \infty} \text{höl}_\beta^*(\tilde{U}_{\pm\tau_n}) = \lim\sup\nolimits_{\tau \to 0} \text{höl}_\beta^*(\tilde{U}_{\pm\tau}) = C_5.$$

W.l.o.g. we can assume that $\text{höl}_\beta^*(\tilde{U}_{\pm\tau_n}) > \frac{C_5}{3}$ for all $n \in \mathbb{N}$. Then we find for every $n \in \mathbb{N}$ a $(x_{\tau_n}, y_{\tau_n}) \in \partial\Sigma \times \partial\Sigma$ such that

$$\frac{|\tilde{U}_{\pm\tau_n}(x_{\tau_n}) - \tilde{U}_{\pm\tau_n}(y_{\tau_n})|}{|x_{\tau_n} - y_{\tau_n}|^\beta} > \frac{C_5}{2}.$$

Since $\partial\Sigma \times \partial\Sigma$ is closed and bounded we can drop to a subsequence $(x_{\tau_{n_k}}, y_{\tau_{n_k}})_{k \in \mathbb{N}}$ converging to $(\bar{x}, \bar{y}) \in \partial\Sigma \times \partial\Sigma$ and a corresponding subsequence $(\tau_{n_k})_{k \in \mathbb{N}}$ still converging to zero. We have to distinguish two different cases. In the first case we have $\bar{x} \neq \bar{y}$. Then we are able to find $K \in \mathbb{N}$ such that $|x_{\tau_{n_k}} - y_{\tau_{n_k}}| > \frac{|\bar{x} - \bar{y}|}{2} =: C_6$ for all $k > K$. Now we can estimate

$$\frac{|\tilde{U}_{\pm\tau_{n_k}}(x_{\tau_{n_k}}) - \tilde{U}_{\pm\tau_{n_k}}(y_{\tau_{n_k}})|}{|x_{\tau_{n_k}} - y_{\tau_{n_k}}|^\beta}$$

$$= \frac{|(\tilde{U}(x_{\tau_{n_k}}) - \tilde{U}(x_{\tau_{n_k}} \pm \tau_{n_k}\nu(x_{\tau_{n_k}}))) - (\tilde{U}(y_{\tau_{n_k}}) + \tilde{U}(y_{\tau_{n_k}} \pm \tau_{n_k}\nu(y_{\tau_{n_k}})))|}{|x_{\tau_{n_k}} - y_{\tau_{n_k}}|^\beta}$$

$$\leq \frac{|(\tilde{U}(x_{\tau_{n_k}}) - \tilde{U}(x_{\tau_{n_k}} \pm \tau_{n_k}\nu(x_{\tau_{n_k}})))| + |(\tilde{U}(y_{\tau_{n_k}}) - \tilde{U}(y_{\tau_{n_k}} \pm \tau_{n_k}\nu(y_{\tau_{n_k}})))|}{C_6^\beta} \leq \frac{2C_7 \tau_{n_k}^\gamma}{C_6^\beta},$$

where C_7 is the Hölder constant for $\tilde{U} \in C^{0,\beta}(\overline{D})$ or $\tilde{U} \in C^{0,\beta}(\overline{\Sigma})$, respectively. But now we have that

$$\frac{|\tilde{U}_{\pm\tau_{n_k}}(x_{\tau_{n_k}}) - \tilde{U}_{\pm\tau_{n_k}}(y_{\tau_{n_k}})|}{|x_{\tau_{n_k}} - y_{\tau_{n_k}}|^\beta} \leq \frac{2C_7 \tau_{n_k}^\gamma}{C_6^\beta} \to 0,$$

for k tending to infinity. This contradicts to

$$\frac{|\tilde{U}_{\pm\tau_n}(x_{\tau_n}) - \tilde{U}_{\pm\tau_n}(y_{\tau_n})|}{|x_{\tau_n} - y_{\tau_n}|^\beta} > \frac{C_5}{2},$$

for all $n \in \mathbb{N}$, and so this case can not occur. So it is left to investigate the second case where $\bar{x} = \bar{y}$. Here we estimate in the following way

$$\frac{|\tilde{U}_{\pm\tau_{n_k}}(x_{\tau_{n_k}}) - \tilde{U}_{\pm\tau_{n_k}}(y_{\tau_{n_k}})|}{|x_{\tau_{n_k}} - y_{\tau_{n_k}}|^\beta} = |x_{\tau_{n_k}} - y_{\tau_{n_k}}|^{\gamma-\beta} \frac{|\tilde{U}_{\pm\tau_{n_k}}(x_{\tau_{n_k}}) - \tilde{U}_{\pm\tau_{n_k}}(y_{\tau_{n_k}})|}{|x_{\tau_{n_k}} - y_{\tau_{n_k}}|^{\gamma-\beta}|x_{\tau_{n_k}} - y_{\tau_{n_k}}|^\beta}$$

$$= |x_{\tau_{n_k}} - y_{\tau_{n_k}}|^{\gamma-\beta} \frac{|\tilde{U}_{\pm\tau_{n_k}}(x_{\tau_{n_k}}) - \tilde{U}_{\pm\tau_{n_k}}(y_{\tau_{n_k}})|}{|x_{\tau_{n_k}} - y_{\tau_{n_k}}|^{\gamma}}$$

$$\leq |x_{\tau_{n_k}} - y_{\tau_{n_k}}|^{\gamma-\beta} 2C_8,$$

where C_8 is the Hölder constant for $\tilde{U} \in C^{0,\gamma}(\overline{D})$ or $\tilde{U} \in C^{0,\gamma}(\overline{\Sigma})$, with $\beta < \gamma < \alpha$. We have

$$\lim_{k \to \infty} |x_{\tau_{n_k}} - y_{\tau_{n_k}}| = 0.$$

Since $\gamma - \beta > 0$ this yields

$$\frac{|\tilde{U}_{\pm\tau_{n_k}}(x_{\tau_{n_k}}) - \tilde{U}_{\pm\tau_{n_k}}(y_{\tau_{n_k}})|}{|x_{\tau_{n_k}} - y_{\tau_{n_k}}|^{\beta}} \leq |x_{\tau_{n_k}} - y_{\tau_{n_k}}|^{\gamma-\beta} 2C_7 \to 0,$$

for k tending to infinity. Like above this is a contradiction and since one of this two cases has to occur our assumption must be wrong and the proof is done. □

Remark 4.4.2. A first result about the convergence in Hölder norms is contained in [FK80] or [FM03]. Here the convergence in $C^{0,\beta}(\partial\Sigma)$ for $F \in C^{0,\alpha}(\partial\Sigma)$ and $\partial\Sigma$ an outer $C^{2,\alpha}$-domain is already proved.

4.5 Limit Formulae in $L^2(\partial\Sigma)$

In this section we present an important result published in [Ker80] and [Fre80] as well as some consequences. The convergence prove in $L^2(\partial\Sigma)$ norm has its origin in [Geh70]. Lemma 4.5.1 contains the convergence of the limit formulae for $U_1[F]$, $U_2[F]$ and $\frac{\partial U_1}{\partial \nu}[F]$ in $L^2(\partial\Sigma)$ norm and is, together with the results of convergence in $C^m(\partial\Sigma)$ norm, the basis for the results in the Sobolev spaces $H^{m,2}(\partial\Sigma)$. The result about the adjoint operators are essential tools for the applications in Section 4.7. We state the result about the convergence in $L^2(\partial\Sigma)$ which is taken from e.g. [Ker80].

Lemma 4.5.1. Let Σ be an outer C^2-domain and $F \in L^2(\partial\Sigma)$. We have $L_i^{\pm\tau}[F] \in L^2(\partial\Sigma)$ for all $\tau \in (0, \tau_0]$, $i = 1, 2, 3$, with

$$\lim_{\tau \to 0^+} \|L_i^{\pm\tau}[F]\|_{L^2(\partial\Sigma)} = 0. \qquad (4.12)$$

Proof. This statement is proved in e.g. [Ker80]. □

4.5. LIMIT FORMULAE IN $L^2(\partial\Sigma)$

Now we present a direct consequence of this lemma which will be one of the main tools in the proof of our main theorem in Section 4.6.

Lemma 4.5.2. Let $\Sigma \subset \mathbb{R}^3$ be an outer C^2-domain and $F \in L^2(\partial\Sigma)$. Then we have

$$\|U_1[F](\,\cdot\, \pm \tau\nu(\,\cdot\,))\|_{L^2(\partial\Sigma)} \leq C_9 \|F\|_{L^2(\partial\Sigma)},$$
$$\|U_2[F](\,\cdot\, \pm \tau\nu(\,\cdot\,))\|_{L^2(\partial\Sigma)} \leq C_{10} \|F\|_{L^2(\partial\Sigma)},$$

where $0 < C_9, C_{10} < \infty$ are constants independent of $\tau \in [0, \tau_0]$. Furthermore, let $(U_1^{\pm\tau})^*$ and $\left(\frac{\partial U_1}{\partial \nu}^{\pm\tau}\right)^*$ be defined by

$$\int_{\partial\Sigma} (U_1^{\pm\tau})^*[F](y) \cdot G(y)\, dH^2(y) = \int_{\partial\Sigma} F(y) \cdot U_1[G](y \pm \tau\nu(y))\, dH^2(y),$$
$$\int_{\partial\Sigma} \left(\frac{\partial U_1}{\partial \nu}^{\pm\tau}\right)^*[F](y) \cdot G(y \pm \tau\nu(y))\, dH^2(y) = \int_{\partial\Sigma} F(y) \cdot \frac{\partial U_1}{\partial \nu}[G](y)\, dH^2(y),$$

for all $\tau \in [0, \tau_0]$, $F, G \in L^2(\partial\Sigma)$, i.e. the Hilbert space adjoint operators in $L^2(\partial\Sigma)$. We have

$$(U_1^{\pm\tau})^*[F](y) = \int_{\partial\Sigma} F(x) \frac{1}{|x \pm \tau\nu(x) - y|}\, dH^2(x),$$
$$\left(\frac{\partial U_1}{\partial \nu}^{\pm\tau}\right)^*[F](y) = \int_{\partial\Sigma} F(x) \frac{\partial}{\partial\nu(z)} \frac{1}{|z - y|}\bigg|_{z = x \pm \tau\nu(x)}\, dH^2(x),$$

for all $y \in \partial\Sigma$, $\tau \in [0, \tau_0]$, $F \in L^2(\partial\Sigma)$, and

$$\|(U_1^{\pm\tau})^*[F]\|_{L^2(\partial\Sigma)} \leq C_{11} \|F\|_{L^2(\partial\Sigma)},$$
$$\left\|\left(\frac{\partial U_1}{\partial \nu}^{\pm\tau}\right)^*[F]\right\|_{L^2(\partial\Sigma)} \leq C_{12} \|F\|_{L^2(\partial\Sigma)},$$

for all $F \in L^2(\partial\Sigma)$ and constants $0 < C_{11}, C_{12} < \infty$ independent of $\tau \in [0, \tau_0]$.

Proof. A byproduct of the proof of Lemma 4.5.1 is

$$\|L_i^{\pm\tau}[F]\|_{L^2(\partial\Sigma)} \leq C_{13} \|F\|_{L^2(\partial\Sigma)},$$

for all $F \in L^2(\partial\Sigma)$, $i = 1, 2, 3$, with a constant $0 < C_{13} < \infty$ independent of $\tau \in (0, \tau_0]$, see e.g. [FK80] or [18, Theorem 3.1], as a result of the uniform boundedness principle. We easily find $\|U_1[F](\,\cdot\, \pm \tau_0\nu(\,\cdot\,))\|_{L^2(\partial\Sigma)} \leq C_{14} \|F\|_{L^2(\partial\Sigma)}$ for $F \in L^2(\partial\Sigma)$ and so

$$\|U_1[F]\|_{L^2(\partial\Sigma)} = \|U_1[F] + U_1[F](\,\cdot\, \pm \tau_0\nu(\,\cdot\,)) - U_1[F](\,\cdot\, \pm \tau_0\nu(\,\cdot\,))\|_{L^2(\partial\Sigma)}$$

$$\leq \|L_1^{\pm \tau_0}[F]\|_{L^2(\partial\Sigma)} + \|U_1[F](\,\cdot\, \pm \tau_0 \nu(\,\cdot\,)\|_{L^2(\partial\Sigma)} \leq (C_{13} + C_{14})\|F\|_{L^2(\partial\Sigma)}$$

for all $F \in L^2(\partial\Sigma)$. Applying the triangle inequality one more time, we gain

$$\|U_1[F](\,\cdot\, \pm \tau\nu(\,\cdot\,)\|_{L^2(\partial\Sigma)} = \|U_1[F](\,\cdot\, \pm \tau\nu(\,\cdot\,) - U_1[F] + U_1[F]\|_{L^2(\partial\Sigma)}$$
$$\leq \|L_1^{\pm\tau}[F]\|_{L^2(\partial\Sigma)} + \|U_1[F]\|_{L^2(\partial\Sigma)} \leq (2C_{13} + C_{14})\|F\|_{L^2(\partial\Sigma)} = C_9\|F\|_{L^2(\partial\Sigma)},$$

for all $F \in L^2(\partial\Sigma)$ with $0 < C_9 < \infty$ independent of $\tau \in (0, \tau_0]$. Similar one can derive the estimate for $U_2[F]$. The formulae of the adjoint operators follow by elementary calculations c.f. [FM04]. Moreover, in [Ker80] Lemma 4.5.1 is also proved for the adjoint operators $\left(L_i^{\pm\tau}\right)^*$, $i = 1, 2, 3$, $\tau \in (0, \tau_0]$, with an analogous estimate for $\|\left(L_i^{\pm\tau}\right)^*[F]\|_{L^2(\partial\Sigma)}$. So the remaining part of the proof can also be done by using the arguments from above and we are done. □

Although we will not use it, we want to mention that such an estimate also exists for $\frac{\partial U_1}{\partial \nu}$ and $(U_2)^*$. It can be proved analogous to Lemma 4.5.2, using the results from [Ker80].

4.6 Limit Formulae in Sobolev Spaces

This section contains the main result of this chapter. We prove the convergence of the limit formulae and jump relations in the Hilbert spaces $H^{m,2}(\partial\Sigma)$, $m \in \mathbb{N}$. Therefore we have to investigate the operators $L_i^{\pm\tau}$, defined in Lemma 4.2.3. In order to do this we will need the results from the previous sections. The following theorem is the main theorem of this section.

Theorem 4.6.1. *Let $m \in \mathbb{N}$, $m \geq 1$, $\alpha \in (0, 1]$ and*

 Σ *be an outer $C^{m+1,\alpha}$-domain and $F \in H^{m,2}(\partial\Sigma)$ for $i = 1$,*

 Σ *be an outer C^{m+2}-domain and $F \in H^{m+1,2}(\partial\Sigma)$ for $i = 2, 3$,*

 Σ *be an outer C^{m+3}-domain and $F \in H^{m+2,2}(\partial\Sigma)$ for $i = 4$.*

Then the operators $L_i^{\pm\tau}$ map to $H^{m,2}(\partial\Sigma)$ for $i = 1, 2, 3, 4$ and all $\tau \in (0, \tau_0]$. Furthermore

$$\lim_{\tau \to 0^+} \|L_i^{\pm\tau}[F]\|_{H^{m,2}(\partial\Sigma)} = 0. \tag{4.13}$$

Moreover, let Σ be an outer C^2-domain and $F \in H^{2,2}(\partial\Sigma)$ for $i = 4$ and $m = 0$. Then the operators $L_4^{\pm\tau}$ map to $L^2(\partial\Sigma)$ for all $\tau \in (0, \tau_0]$ and

$$\lim_{\tau \to 0^+} \|L_i^{\pm\tau}[F]\|_{L^2(\partial\Sigma)} = 0. \tag{4.14}$$

4.6. LIMIT FORMULAE IN SOBOLEV SPACES

Proof. At first we prove the theorem for $L_3^{\pm\tau}$. Let $F \in C^{m+1}(\partial\Sigma)$ be given. We will show that

$$\|U_2[F](\,\cdot\, \pm \tau\nu(\,\cdot\,))\|_{H^{m,2}(\partial\Sigma)} \leq C_{15}\|F\|_{H^{m+1,2}(\partial\Sigma)},$$

where $0 < C_{15} < \infty$ is a constant independent of $\tau \in (0, \tau_0]$. Then we have

$$\|U_2[F](\,\cdot\,)\|_{H^{m,2}(\partial\Sigma)}$$
$$=\|U_2[F](\,\cdot\,) \pm 2\pi F - U_2[F](\,\cdot\, \pm \tau\nu(\,\cdot\,)) \mp 2\pi F + U_2[F](\,\cdot\, \pm \tau\nu(\,\cdot\,))\|_{H^{m,2}(\partial\Sigma)}$$
$$\leq \|L_3^{\pm\tau}[F]\|_{H^{m,2}(\partial\Sigma)} + 2\pi\|F\|_{H^{m,2}(\partial\Sigma)} + \|U_2[F](\,\cdot\, \pm \tau\nu(\,\cdot\,))\|_{H^{m,2}(\partial\Sigma)}$$
$$\leq H^2(\partial\Sigma)\|L_3^{\pm\tau}[F]\|_{C^m(\partial\Sigma)} + (C_{15} + 2\pi)\|F\|_{H^{m+1,2}(\partial\Sigma)},$$

for all $F \in C^{m+1}(\partial\Sigma)$. Since this has to hold for all $\tau \in (0, \tau_0]$, $C^{m+1}(\partial\Sigma) \subset C^{m,\alpha}(\partial\Sigma)$ and each outer C^{m+2}-domain is also an outer $C^{m+1,\beta}$-domain, $\beta \in [0,1]$, Theorem 4.3.3 yields

$$\|U_2[F](\,\cdot\,)\|_{H^{m,2}(\partial\Sigma)} \leq (C_{15} + 2\pi)\|F\|_{H^{m+1,2}(\partial\Sigma)}$$

and consequently

$$\|L_3^{\pm\tau}[F]\|_{H^{m,2}(\partial\Sigma)} \leq (2C_{15} + 2\pi + 1)\|F\|_{H^{m+1,2}(\partial\Sigma)},$$

for all $F \in C^{m+1}(\partial\Sigma)$. Thus, the BLT Theorem, see Lemma 2.4.5, gives the unique linear continuation

$$L_3^{\pm\tau} : H^{m+1,2}(\partial\Sigma) \to H^{m,2}(\partial\Sigma),$$

with same bound. Finally, we have for arbitrary $F \in H^{m+1,2}(\partial\Sigma)$ and a sequence $(F_n)_{n\in\mathbb{N}} \subset C^{m+1}(\partial\Sigma)$ such that $F_n \to F$ in $H^{m+1,2}(\partial\Sigma)$

$$\lim_{\tau \to 0+} \|L_3^{\pm\tau}[F]\|_{H^{m,2}(\partial\Sigma)} = \lim_{\tau \to 0+} \|L_3^{\pm\tau}[F] - L_3^{\pm\tau}[F_n] + L_3^{\pm\tau}[F_n]\|_{H^{m,2}(\partial\Sigma)}$$
$$\leq \lim_{\tau \to 0+} \|L_3^{\pm\tau}[F - F_n]\|_{H^{m,2}(\partial\Sigma)} + \|L_3^{\pm\tau}[F_n]\|_{H^{m,2}(\partial\Sigma)}$$
$$\leq \lim_{\tau \to 0+} 2(C_{15} + 2\pi)\|F - F_n\|_{H^{m+1,2}(\partial\Sigma)} + H^2(\partial\Sigma)\|L_3^{\pm\tau}[F_n]\|_{C^m(\partial\Sigma)}$$
$$= 2(C_{15} + 2\pi)\|F - F_n\|_{H^{m+1,2}(\partial\Sigma)}.$$

Because this holds for all $n \in \mathbb{N}$, the theorem for $L_3^{\pm\tau}[F]$ is proved. So it is left to show that

$$\|U_2[F](\,\cdot\, \pm \tau\nu(\,\cdot\,))\|_{H^{m,2}(\partial\Sigma)} \leq C_{15}\|F\|_{H^{m+1,2}(\partial\Sigma)},$$

for all $F \in C^{m+1}(\partial\Sigma)$, where $0 < C_{13} < \infty$ is a constant independent of $\tau \in (0, \tau_0]$. We have

$$\|U_2[F](\,\cdot\, \pm \tau\nu(\,\cdot\,))\|^2_{H^{m,2}(\partial\Sigma)}$$

$$= \sum_{i=1}^{N} \|U_2[F](\Psi_i(\,\cdot\,) \pm \tau\nu(\Psi_i(\,\cdot\,)))\|^2_{H^{m,2}(B_1^{\mathbb{R}^2}(0))}$$

$$= \sum_{i=1}^{N} \sum_{s_1+s_2=0}^{m} \|\partial_1^{s_1}\partial_2^{s_2} U_2[F](\Psi_i(\,\cdot\,) \pm \tau\nu(\Psi_i(\,\cdot\,)))\|^2_{L^2(B_1^{\mathbb{R}^2}(0))}$$

$$= \sum_{i=1}^{N} \sum_{s_1+s_2=0}^{m} \|\partial_1^{s_1}\partial_2^{s_2} \int_{\partial\Sigma} F(z) \frac{\partial}{\partial\nu(z)} \frac{1}{|\Psi_i(\,\cdot\,) \pm \tau\nu(\Psi_i(\,\cdot\,)) - z|} dH^2(z)\|^2_{L^2(B_1^{\mathbb{R}^2}(0))}$$

$$= \sum_{i=1}^{N} \sum_{s_1+s_2=0}^{m} \|\partial_1^{s_1}\partial_2^{s_2} \sum_{j=1}^{N} \int_{B_1^{\mathbb{R}^2}(0)} w_j(\Psi_j(y))F(\Psi_j(y)) \tag{4.15}$$

$$\left.\frac{\partial}{\partial\nu(z)} \frac{1}{|\Psi_i(\,\cdot\,) \pm \tau\nu(\Psi_i(\,\cdot\,)) - z|}\right|_{z=\Psi_j(y)} J_j(y)d\lambda^2(y)\|^2_{L^2(B_1^{\mathbb{R}^2}(0))}$$

$$= \sum_{i=1}^{N} \sum_{s_1+s_2=0}^{m} \|\partial_1^{s_1}\partial_2^{s_2} \sum_{j=1}^{N} \int_{B_1^{\mathbb{R}^2}(0)} w_j(\Psi_j(y))F(\Psi_j(y)) \tag{4.16}$$

$$\frac{\partial}{\partial\tilde{\nu}(z)} \frac{1}{|\Psi_i(\,\cdot\,) \pm \tau\nu(\Psi_i(\,\cdot\,)) - \Psi_j(y)|} J_j(y)d\lambda^2(y)\|^2_{L^2(B_1^{\mathbb{R}^2}(0))}$$

$$\leq \sum_{i,j=1}^{N} \sum_{s_1+s_2=0}^{m} \|\partial_1^{s_1}\partial_2^{s_2} \Big(w_i(\Psi_i(\,\cdot\,)) \int_{B_1^{\mathbb{R}^2}(0)} w_j(\Psi_j(y))F(\Psi_j(y)) \tag{4.17}$$

$$\frac{\partial}{\partial\tilde{\nu}(y)} \frac{1}{|\Psi_i(\,\cdot\,) \pm \tau\nu(\Psi_i(\,\cdot\,)) - \Psi_j(y)|} J_j(y)d\lambda^2(y)\Big)\|^2_{L^2(B_1^{\mathbb{R}^2}(0))}$$

$$\leq \sum_{i,j=1}^{N} \sum_{s_1+s_2=0}^{m} \sum_{k_1=0}^{s_1} \sum_{k_2=0}^{s_2} \binom{s_1}{k_1}\binom{s_2}{k_2} \|\partial_1^{s_1-k_1}\partial_2^{s_2-k_2}\big(w_i(\Psi_i(\,\cdot\,))\big) \tag{4.18}$$

$$\partial_1^{k_1}\partial_2^{k_2} \int_{B_1^{\mathbb{R}^2}(0)} w_j(\Psi_j(y))F(\Psi_j(y)) \frac{\partial}{\partial\tilde{\nu}(y)} \frac{1}{|\Psi_i(\,\cdot\,) \pm \tau\nu(\Psi_i(\,\cdot\,)) - \Psi_j(y)|} J_j(y)d\lambda^2(y)\|^2_{L^2(B_1^{\mathbb{R}^2}(0))}$$

$$\leq \sum_{i,j=1}^{N} \sum_{s_1+s_2=0}^{m} \sum_{k_1=0}^{s_1} \sum_{k_2=0}^{s_2} \binom{s_1}{k_1}\binom{s_2}{k_2} \underbrace{\|w_i\|_{C^m(\partial\Sigma)}}_{<\infty}$$

$$\cdot \|\partial_1^{k_1}\partial_2^{k_2} \int_{B_1^{\mathbb{R}^2}(0)} w_j(\Psi_j(y))F(\Psi_j(y)) \frac{\partial}{\partial\tilde{\nu}(y)} \frac{1}{|\Psi_i(\,\cdot\,) \pm \tau\nu(\Psi_i(\,\cdot\,)) - \Psi_j(y)|} J_j(y)d\lambda^2(y)\|^2_{L^2(B_1^{\mathbb{R}^2}(0))} \tag{4.19}$$

by the binomial theorem, where $\Psi_j(y)$ denotes $\Psi_j(y_1, y_2, 0)$. We used that Ψ_j, $j = 1, \ldots, N$ are C^1-diffeomorphisms and

$$\left(\frac{\partial}{\partial\nu(\Psi_j(y))}G\right)(\Psi_j(y)) = DG(\Psi_j(y))\nu(\Psi_j(y)) = D(G(\Psi_j(y)))\underbrace{(D\Psi_j(y))^{-1}\nu(\Psi_j(y))}_{=:\tilde{\nu}(y)}$$

4.6. LIMIT FORMULAE IN SOBOLEV SPACES

for each function G, which is totally differentiable at the point $\Psi_j(y)$. DG means the total differential of G. Of course $\frac{1}{|x \pm \tau \nu(x) - y|}$ is totally differentiable for each $y \in \partial \Sigma$ as long as $\tau \neq 0$. Because for a C^{m+2}-surface $\partial \Sigma$, $\nu \in C^{m+1}(\partial \Sigma)$, see Lemma 2.1.6, and by definition $\Psi_j \in C^{m+2}(\overline{B_1^{\mathbb{R}^3}(0)})$, we consequently have $\tilde{\nu} \in C^{m+1}(\overline{B_1^{\mathbb{R}^2}(0)})$. It is left to estimate the single summands from the sum above. Before we start the estimation, we modify the terms as follows and start with the case of $|s| = 1$, i.e., with terms of the form

$$\partial_k \int_{B_1^{\mathbb{R}^2}(0)} w_j(\Psi_j(y)) F(\Psi_j(y)) \frac{\partial}{\partial \tilde{\nu}(y)} \frac{1}{|\Psi_i(\cdot) \pm \tau \nu(\Psi_i(\cdot)) - \Psi_j(y)|} J_j(y) d\lambda^2(y), \qquad (4.20)$$

for $k \in \{1,2\}$, $i,j \in \{1,\ldots,N\}$, $F \in C^{m+1}(\partial \Sigma)$ and $\tau \in (0, \tau_0]$ arbitrary but fixed. Using Lemma 2.4.8 we can interchange differentiation and integration as well as the order of differentiation to get

$$\frac{\partial}{\partial x_k} \int_{B_1^{\mathbb{R}^2}(0)} w_j(\Psi_j(y)) F(\Psi_j(y)) \frac{\partial}{\partial \tilde{\nu}(y)} \frac{1}{|\Psi_i(x) \pm \tau \nu(\Psi_i(x)) - \Psi_j(y)|} J_j(y) d\lambda^2(y)$$
$$= \int_{B_1^{\mathbb{R}^2}(0)} w_j(\Psi_j(y)) F(\Psi_j(y)) \frac{\partial}{\partial \tilde{\nu}(y)} \frac{\partial}{\partial x_k} \frac{1}{|\Psi_i(x) \pm \tau \nu(\Psi_i(x)) - \Psi_j(y)|} J_j(y) d\lambda^2(y).$$

We will now translate the differentiation with respect to x into a differentiation with respect to y. We have

$$\frac{\partial}{\partial x_k} \frac{1}{|\Psi_i(x) \pm \tau \nu(\Psi_i(x)) - \Psi_j(y)|} = -\frac{\langle \Psi_i(x) \pm \tau \nu(\Psi_i(x)) - \Psi_j(y), \partial_k (\Psi_i(x) \pm \tau \nu(\Psi_i(x))) \rangle}{|\Psi_i(x) \pm \tau \nu(\Psi_i(x)) - \Psi_j(y)|^3}$$

$$\frac{\partial}{\partial y_l} \frac{1}{|\Psi_i(x) \pm \tau \nu(\Psi_i(x)) - \Psi_j(y)|} = \frac{\langle \Psi_i(x) \pm \tau \nu(\Psi_i(x)) - \Psi_j(y), \partial_l \Psi_j(y) \rangle}{|\Psi_i(x) \pm \tau \nu(\Psi_i(x)) - \Psi_j(y)|^3}$$

for $k, l = 1,2,3$. Because $\text{Det}(D\Psi_j(y)) \neq 0$ for all $y \in \overline{B_1^{\mathbb{R}^2}(0)}$ we have that a basis of $\mathbb{R}^3$ given by

$$\{\partial_1 \Psi_j(y), \partial_2 \Psi_j(y), \partial_3 \Psi_j(y)\}$$

for each $y \in B_1^{\mathbb{R}^2}(0)$. Consequently we find uniquely determined functions $a_1^k(x,y)$, $a_2^k(x,y)$ and $a_3^k(x,y)$ such that

$$\frac{\partial}{\partial x_k} \frac{1}{|\Psi_i(x) \pm \tau \nu(\Psi_i(x)) - \Psi_j(y)|} = -a_1^k(x,y) \frac{\partial}{\partial y_1} \frac{1}{|\Psi_i(x) \pm \tau \nu(\Psi_i(x)) - \Psi_j(y)|}$$
$$- a_2^k(x,y) \frac{\partial}{\partial y_2} \frac{1}{|\Psi_i(x) \pm \tau \nu(\Psi_i(x)) - \Psi_j(y)|} - a_3^k(x,y) \frac{\partial}{\partial y_3} \frac{1}{|\Psi_i(x) \pm \tau \nu(\Psi_i(x)) - \Psi_j(y)|},$$

where $a^k(x,y) = (a_1^k(x,y), a_2^k(x,y), a_3^k(x,y))^T$ can be computed by

$$a^k(x,y) = -\big(D\Psi_j(y)\big)^{-1} \partial_k \big(\Psi_i(x) \pm \tau \nu(\Psi_i(x))\big).$$

Moreover, by the inversion formula for 3×3 matrices and the fact that Ψ_j is a C^{m+2}-diffeomorphism, we find the components of the matrix $\left(D\Psi_j(y)\right)^{-1}$ in $C^{m+1}(\overline{B_1^{\mathbb{R}^2}(0)})$. This yields

$$a_l^k(x,y) = b_l^k(y) \cdot c_l^{k,\pm\tau}(x)$$

with $b_l^k, c_l^{k,\pm\tau} \in C^{m+1}(\overline{B_1^{\mathbb{R}^2}(0)})$ for $l = 1, 2, 3$. Additionally we have

$$\|c_l^{k,\pm\tau}\|_{C^{m+1}(\overline{B_1^{\mathbb{R}^2}(0)})} \leq \sum_{p,q=1}^{3} \|(D\Psi_j)_{pq}^{-1}\|_{C^{m+1}(\overline{B_1^{\mathbb{R}^2}(0)})}$$
$$\cdot \left(\|\partial_k \Psi_i(x)\|_{C^{m+1}(\overline{B_1^{\mathbb{R}^2}(0)})} + \tau_0 \|\partial_k \nu(\Psi_i(x))\|_{C^{m+1}(\overline{B_1^{\mathbb{R}^2}(0)})} \right).$$

So although $c_l^{k,\pm\tau}$ depends on τ, its norm is bounded by a constant independent of $\tau \in (0, \tau_0]$. Applying the whole considerations to the term form (4.20), we get

$$\frac{\partial}{\partial x_k} \int_{B_1^{\mathbb{R}^2}(0)} w_j(\Psi_j(y)) F(\Psi_j(y)) \frac{\partial}{\partial \tilde{\nu}(y)} \frac{1}{|\Psi_i(x) \pm \tau\nu(\Psi_i(x)) - \Psi_j(y)|} J_j(y) d\lambda^2(y)$$
$$= -\sum_{n=1}^{3}\sum_{l=1}^{3} \int_{B_1^{\mathbb{R}^2}(0)} w_j(\Psi_j(y)) F(\Psi_j(y)) J_j(y)$$
$$\tilde{\nu}_n(y) \frac{\partial}{\partial y_n}\left(b_l^k(y) c_l^{k,\pm\tau}(x) \frac{\partial}{\partial y_l} \frac{1}{|\Psi_i(x) \pm \tau\nu(\Psi_i(x)) - \Psi_j(y)|}\right) d\lambda^2(y)$$
$$= -\sum_{n=1}^{3}\sum_{l=1}^{3} \int_{B_1^{\mathbb{R}^2}(0)} w_j(\Psi_j(y)) F(\Psi_j(y)) J_j(y) \tilde{\nu}_n(y) b_l^k(y) c_l^{k,\pm\tau}(x)$$
$$\frac{\partial}{\partial y_n}\frac{\partial}{\partial y_l} \frac{1}{|\Psi_i(x) \pm \tau\nu(\Psi_i(x)) - \Psi_j(y)|} d\lambda^2(y)$$
$$-\sum_{n=1}^{3}\sum_{l=1}^{3} \int_{B_1^{\mathbb{R}^2}(0)} w_j(\Psi_j(y)) F(\Psi_j(y)) J_j(y) \left(\frac{\partial}{\partial y_n} b_l^k(y)\right) c_l^{k,\pm\tau}(x) \tilde{\nu}_n(y)$$
$$\frac{\partial}{\partial y_l} \frac{1}{|\Psi_i(x) \pm \tau\nu(\Psi_i(x)) - \Psi_j(y)|} d\lambda^2(y).$$

For $n,l \in \{1,2\}$ we apply integration by parts. Note that because of the C^∞-partition of the unity w_j the boundary integrals disappear. Consequently we get

$$\frac{\partial}{\partial x_k} \int_{B_1^{\mathbb{R}^2}(0)} w_j(\Psi_j(y)) F(\Psi_j(y)) \frac{\partial}{\partial \tilde{\nu}(y)} \frac{1}{|\Psi_i(x) \pm \tau\nu(\Psi_i(x)) - \Psi_j(y)|} J_j(y) d\lambda^2(y)$$
$$= -\int_{B_1^{\mathbb{R}^2}(0)} w_j(\Psi_j(y)) F(\Psi_j(y)) J_j(y) \tilde{\nu}_n(y) b_3^3(x) c_3^{3,\pm\tau}(y)$$

$$\frac{\partial^2}{\partial y_3^2} \frac{1}{|\Psi_i(x) \pm \tau\nu(\Psi_i(x)) - \Psi_j(y)|} d\lambda^2(y) \qquad (4.21)$$

$$+ \sum_{n=1}^{2} \int_{B_1^{\mathbb{R}^2}(0)} \frac{\partial}{\partial y_n} \left(w_j(\Psi_j(y)) F(\Psi_j(y)) J_j(y) \tilde{\nu}_n(y) b_3^k(x) c_3^{k,\pm\tau}(y) \right)$$

$$\frac{\partial}{\partial y_3} \frac{1}{|\Psi_i(x) \pm \tau\nu(\Psi_i(x)) - \Psi_j(y)|} d\lambda^2(y) \qquad (4.22)$$

$$+ \sum_{l=1}^{2} \int_{B_1^{\mathbb{R}^2}(0)} \frac{\partial}{\partial y_l} \left(w_j(\Psi_j(y)) F(\Psi_j(y)) J_j(y) \tilde{\nu}_3(y) b_l^k(x) c_l^{k,\pm\tau}(y) \right)$$

$$\frac{\partial}{\partial y_3} \frac{1}{|\Psi_i(x) \pm \tau\nu(\Psi_i(x)) - \Psi_j(y)|} d\lambda^2(y) \qquad (4.23)$$

$$- \sum_{n=1}^{2} \sum_{l=1}^{2} \int_{B_1^{\mathbb{R}^2}(0)} \frac{\partial}{\partial y_n} \frac{\partial}{\partial y_l} \left(w_j(\Psi_j(y)) F(\Psi_j(y)) J_j(y) \tilde{\nu}_n(y) b_l^k(x) c_l^{k,\pm\tau}(y) \right)$$

$$\frac{1}{|\Psi_i(x) \pm \tau\nu(\Psi_i(x)) - \Psi_j(y)|} d\lambda^2(y) \qquad (4.24)$$

$$+ \sum_{n=1}^{3} \sum_{l=1}^{2} \int_{B_1^{\mathbb{R}^2}(0)} \frac{\partial}{\partial y_l} \left(w_j(\Psi_j(y)) F(\Psi_j(y)) J_j(y) b_l^k(x) \left(\frac{\partial}{\partial y_n} c_l^{k,\pm\tau}(y) \right) \tilde{\nu}_n(y) \right)$$

$$\frac{1}{|\Psi_i(x) \pm \tau\nu(\Psi_i(x)) - \Psi_j(y)|} d\lambda^2(y) \qquad (4.25)$$

$$- \sum_{n=1}^{3} \int_{B_1^{\mathbb{R}^2}(0)} w_j(\Psi_j(y)) F(\Psi_j(y)) J_j(y) \left(\frac{\partial}{\partial y_n} b_3^k(x) c_3^{k,\pm\tau}(y) \right) \tilde{\nu}_n(y)$$

$$\frac{\partial}{\partial y_3} \frac{1}{|\Psi_i(x) \pm \tau\nu(\Psi_i(x)) - \Psi_j(y)|} d\lambda^2(y). \qquad (1.26)$$

Before we go on, we modify (4.21) as follows. Consider the term

$$\int_{B_1^{\mathbb{R}^2}(0)} w_j(\Psi_j(y)) F(\Psi_j(y)) J_j(y) \tilde{\nu}_n(y) b_3^3(x) c_3^{3,\pm\tau}(y)$$

$$\frac{\partial^2}{\partial y_3^2} \frac{1}{|\Psi_i(x) \pm \tau\nu(\Psi_i(x)) - \Psi_j(y)|} d\lambda^2(y).$$

We use the following transformation formula for the Laplace operator for twice continuously differentiable functions H and C^2-diffeomorphisms Ψ, see e.g. [For07],

$$(\Delta H)(\Psi(x)) = \left(\frac{1}{\sqrt{g}} \sum_{i,j=1}^{3} \frac{\partial}{\partial y_j} \left(g^{ij} \sqrt{g} \frac{\partial}{\partial y_i} \right) \right) (H(\Psi(x))).$$

This yields for harmonic functions

$$0 = \left(\frac{1}{\sqrt{g}} \sum_{k,j=1}^{3} \frac{\partial}{\partial y_j} \left(g^{kj} \sqrt{g} \frac{\partial}{\partial y_k} \right) \right) (H(\Psi(x))),$$

where g and g^{kj} are the well known differential geometric functions independent of $\tau \in (0, \tau_0]$. They are defined by

$$\left(g_{kj}\right)_{k,j=1,2,3}(x) := \langle \partial_k \Psi(x), \partial_j \Psi(x) \rangle,$$

$$g(x) := \mathrm{Det}\left(\left((g_{kj})_{k,j=1,2,3}(x)\right)\right),$$

$$\left(g^{kj}\right)_{k,j=1,2,3}(x) := \left(g_{kj}(x)\right)^{-1}_{k,j=1,2,3},$$

for all $x \in B_1^{\mathbb{R}^3}(0)$. For details see [22, pp. 28]. All we need to know in the following is that $g, g^{kj} \in C^{m+1}(\overline{B_1^{\mathbb{R}^3}(0)})$, $k, j = 1, 2, 3$, if we choose $\Psi = \Psi_i$, $i \in \{1, \ldots, N\}$, for a C^{m+2}-surface $\partial\Sigma$, because they are defined with help of the first order derivatives of the mapping Ψ. Additionally we have $g^{33} \neq 0$ on $\overline{B_1^{\mathbb{R}^2}(0)} \times \{0\}$, because by the inversion formula for invertible 3×3 matrices we have

$$g^{33}(x) = \frac{\mathrm{Det}\left(\left((g_{kj})_{k,j=1,2}(x)\right)\right)}{\mathrm{Det}\left(\left((g_{kj})_{k,j=1,2,3}(x)\right)\right)},$$

for all $x \in B_1^{\mathbb{R}^2}(0)$, and $g(x) = \mathrm{Det}(D\Psi_i(x)) > 0$ on $\overline{B_1^{\mathbb{R}^2}(0)}$. Furthermore, [7, Section 2.1] gives

$$\mathrm{Det}\left(\left(g_{kj}\right)_{k,j=1,2}(x)\right) > 0$$

on $\overline{B_1^{\mathbb{R}^2}(0)}$. So rearranging the equation above with respect to $\frac{\partial^2}{\partial y_3^2}$ we get

$$-\frac{\partial^2}{\partial y_3^2}(H((\Psi)))$$

$$= \left(\frac{1}{g^{33}\sqrt{g}}\left(\sum_{k,j=1}^{2} \frac{\partial}{\partial y_j}\left(g^{kj}\sqrt{g}\right)\frac{\partial}{\partial y_k} + \sum_{k=1}^{2} \frac{\partial}{\partial y_3}\left(g^{k3}\sqrt{g}\right)\frac{\partial}{\partial y_k}\right)\right)(H(\Psi(x)))$$

$$+ \left(\frac{1}{g^{33}\sqrt{g}}\sum_{k,j=1}^{2} g^{kj}\sqrt{g}\frac{\partial}{\partial y_j}\frac{\partial}{\partial y_k}\right)(H(\Psi(x)))$$

$$+ \left(\frac{1}{g^{33}\sqrt{g}}\left(\sum_{k=1}^{2} g^{k3}\sqrt{g}\frac{\partial}{\partial y_k} + \sum_{j=1}^{2} \frac{\partial}{\partial y_j}\left(g^{3j}\sqrt{g}\right)\right)\right)\frac{\partial}{\partial y_3}(H(\Psi(x)))$$

$$+ \left(\frac{1}{g^{33}\sqrt{g}}\left(\frac{\partial}{\partial y_3}\left(g^{33}\sqrt{g}\right) + \sum_{j=1}^{2} g^{3j}\sqrt{g}\frac{\partial}{\partial y_j}\right)\right)\frac{\partial}{\partial y_3}(H(\Psi(x))).$$

Applying this formula to the term given by (4.21) and using integration by parts for $k, l = 1, 2$, we end up with the following terms

$$-\int_{B_1^{\mathbb{R}^2}(0)} w_j(\Psi_j(y)) F(\Psi_j(y)) J_j(y) \tilde{\nu}_n(y) b_3^3(x) c_3^{3,\pm\tau}(y)$$

4.6. LIMIT FORMULAE IN SOBOLEV SPACES

$$\frac{\partial^2}{\partial y_3^2} \frac{1}{|\Psi_i(x) \pm \tau\nu(\Psi_i(x)) - \Psi_j(y)|} d\lambda^2(y)$$

$$= \sum_{k,l=1}^{2} \int_{B_1^{\mathbb{R}^2}(0)} w_j(\Psi_j(y)) F(\Psi_j(y)) J_j(y) \tilde{\nu}_n(y) b_3^3(x) c_3^{3,\pm\tau}(y)$$

$$\frac{1}{g^{33}(y)\sqrt{g(y)}} \frac{\partial}{\partial y_l} \left(g^{kl}(y) \sqrt{g(y)} \right) \frac{\partial}{\partial y_k} \frac{1}{|\Psi_i(x) \pm \tau\nu(\Psi_i(x)) - \Psi_j(y)|} d\lambda^2(y) \tag{4.27}$$

$$+ \sum_{k=1}^{2} \int_{B_1^{\mathbb{R}^2}(0)} w_j(\Psi_j(y)) F(\Psi_j(y)) J_j(y) \tilde{\nu}_n(y) b_3^3(x) c_3^{3,\pm\tau}(y)$$

$$\frac{1}{g^{33}(y)\sqrt{g(y)}} \frac{\partial}{\partial y_3} \left(g^{k3}(y) \sqrt{g(y)} \right) \frac{\partial}{\partial y_k} \frac{1}{|\Psi_i(x) \pm \tau\nu(\Psi_i(x)) - \Psi_j(y)|} d\lambda^2(y) \tag{4.28}$$

$$+ \sum_{k,l=1}^{2} \int_{B_1^{\mathbb{R}^2}(0)} w_j(\Psi_j(y)) F(\Psi_j(y)) J_j(y) \tilde{\nu}_n(y) b_3^3(x) c_3^{3,\pm\tau}(y)$$

$$\frac{1}{g^{33}(y)\sqrt{g(y)}} \left(g^{kl}(y) \sqrt{g(y)} \right) \frac{\partial}{\partial y_k} \frac{\partial}{\partial y_l} \frac{1}{|\Psi_i(x) \pm \tau\nu(\Psi_i(x)) - \Psi_j(y)|} d\lambda^2(y) \tag{4.29}$$

$$+ \sum_{k=1}^{2} \int_{B_1^{\mathbb{R}^2}(0)} w_j(\Psi_j(y)) F(\Psi_j(y)) J_j(y) \tilde{\nu}_n(y) b_3^3(x) c_3^{3,\pm\tau}(y)$$

$$\frac{1}{g^{33}(y)\sqrt{g(y)}} \left(g^{k3}(y) \sqrt{g(y)} \right) \frac{\partial}{\partial y_k} \frac{\partial}{\partial y_3} \frac{1}{|\Psi_i(x) \pm \tau\nu(\Psi_i(x)) - \Psi_j(y)|} d\lambda^2(y) \tag{4.30}$$

$$+ \sum_{k=1}^{2} \int_{B_1^{\mathbb{R}^2}(0)} w_j(\Psi_j(y)) F(\Psi_j(y)) J_j(y) \tilde{\nu}_n(y) b_3^3(x) c_3^{3,\pm\tau}(y)$$

$$\frac{1}{g^{33}(y)\sqrt{g(y)}} \frac{\partial}{\partial y_k} \left(g^{3k}(y) \sqrt{g(y)} \right) \frac{\partial}{\partial y_3} \frac{1}{|\Psi_i(x) \pm \tau\nu(\Psi_i(x)) - \Psi_j(y)|} d\lambda^2(y) \tag{4.31}$$

$$+ \sum_{k=1}^{2} \int_{B_1^{\mathbb{R}^2}(0)} w_j(\Psi_j(y)) F(\Psi_j(y)) J_j(y) \tilde{\nu}_n(y) b_3^3(x) c_3^{3,\pm\tau}(y)$$

$$\frac{1}{g^{33}(y)\sqrt{g(y)}} \left(g^{3k}(y) \sqrt{g(y)} \right) \frac{\partial}{\partial y_k} \frac{\partial}{\partial y_3} \frac{1}{|\Psi_i(x) \pm \tau\nu(\Psi_i(x)) - \Psi_j(y)|} d\lambda^2(y) \tag{4.32}$$

$$+ \int_{B_1^{\mathbb{R}^2}(0)} w_j(\Psi_j(y)) F(\Psi_j(y)) J_j(y) \tilde{\nu}_n(y) b_3^3(x) c_3^{3,\pm\tau}(y)$$
$$\frac{1}{g^{33}(y)\sqrt{g(y)}} \frac{\partial}{\partial y_3} \left(g^{33}(y)\sqrt{g(y)} \right) \frac{\partial}{\partial y_3} \frac{1}{|\Psi_i(x) \pm \tau\nu(\Psi_i(x)) - \Psi_j(y)|} d\lambda^2(y) \quad (4.33)$$

Taking a closer look on (4.22)-(4.26) as well as (4.27)-(4.33), we find that all terms are covered by one of the following two cases

$$\int_{B_1^{\mathbb{R}^2}(0)} \partial_1^{r_1} \partial_2^{r_2} \left(F(\Psi_j(y)) \right) G_1(y) H_1(x) \frac{1}{|\Psi_i(x) \pm \tau\nu(\Psi_i(x)) - \Psi_j(y)|} d\lambda^2(y), \quad (4.34)$$

$$\int_{B_1^{\mathbb{R}^2}(0)} \partial_1^{t_1} \partial_2^{t_2} \left(F(\Psi_j(y)) \right) G_2(y) H_2(x) \frac{\partial}{\partial y_3} \frac{1}{|\Psi_i(x) \pm \tau\nu(\Psi_i(x)) - \Psi_j(y)|} d\lambda^2(y), \quad (4.35)$$

with $0 \leq r_1 + r_2 \leq 2$, $0 \leq t_1 + t_2 \leq 1$, $G_1 \in C^{m-1}\left(\overline{B_1^{\mathbb{R}^2}(0)}\right)$, $G_2 \in C^m\left(\overline{B_1^{\mathbb{R}^2}(0)}\right)$ and $H_1, H_2 \in C^{m+1}\left(\overline{B_1^{\mathbb{R}^2}(0)}\right)$. For arbitrary $1 \leq |s| \leq m$ we now iterate the techniques presented above and end up with the following types of terms

$$\int_{B_1^{\mathbb{R}^2}(0)} \partial_1^{r_1} \partial_2^{r_2} \left(F(\Psi_j(y)) \right) G_1(y) H_1(x) \frac{1}{|\Psi_i(x) \pm \tau\nu(\Psi_i(x)) - \Psi_j(y)|} d\lambda^2(y), \quad (4.36)$$

$$\int_{B_1^{\mathbb{R}^2}(0)} \partial_1^{t_1} \partial_2^{t_2} \left(F(\Psi_j(y)) \right) G_2(y) H_2(x) \frac{\partial}{\partial y_3} \frac{1}{|\Psi_i(x) \pm \tau\nu(\Psi_i(x)) - \Psi_j(y)|} d\lambda^2(y), \quad (4.37)$$

with $0 \leq r_1 + r_2 \leq m+1$, $0 \leq t_1 + t_2 \leq m$, $G_1 \in C^0(\overline{B_1^{\mathbb{R}^2}(0)})$, $G_2 \in C^1\left(\overline{B_1^{\mathbb{R}^2}(0)}\right)$ and $H_1, H_2 \in C^2\left(\overline{B_1^{\mathbb{R}^2}(0)}\right)$. In order to reduce term (4.37) to the case of a double layer potential we use

$$\frac{\partial}{\partial \tilde{\nu}(y)} = \sum_{q=1}^{3} \tilde{\nu}_q(y) \frac{\partial}{\partial y_q}$$

and consequently

$$\frac{\partial}{\partial y_3} = \frac{1}{\tilde{\nu}_3(y)} \frac{\partial}{\partial \tilde{\nu}(y)} - \sum_{i=q}^{2} \frac{\tilde{\nu}_q(y)}{\tilde{\nu}_3(y)} \frac{\partial}{\partial y_q}$$

for each function which is totally differentiable in $(y, 0) \in B_1^{\mathbb{R}^3}(0)$. Recall that the total differentiability is given for all $\tau \in (0, \tau_0]$ as well as the definition of $\tilde{\nu}$. Since $\partial_1 \Psi_j$ and $\partial_2 \Psi_j$ span the tangential space $T_{\partial\Sigma}(y)$ and ν is, as the normal vector, not an element of that space we

4.6. LIMIT FORMULAE IN SOBOLEV SPACES

have that $\tilde{\nu}_3(y) \neq 0$ for all $y \in \overline{B_1^{\mathbb{R}^2}(0)}$, because $\nu(\Psi_j(y)) = D\Psi_j(y)\tilde{\nu}(y)$ on $\overline{B_1^{\mathbb{R}^2}(0)}$. This now yields

$$\int_{B_1^{\mathbb{R}^2}(0)} \partial_1^{t_1} \partial_2^{t_2} \left(F(\Psi_j(y))\right) G_2(y) H_2(x) \frac{\partial}{\partial y_3} \frac{1}{|\Psi_i(x) \pm \tau\nu(\Psi_i(x)) - \Psi_j(y)|} d\lambda^2(y)$$

$$= \int_{B_1^{\mathbb{R}^2}(0)} \partial_1^{t_1} \partial_2^{t_2} \left(F(\Psi_j(y))\right) G_2(y) H_2(x) \frac{1}{\tilde{\nu}_3(y)} \frac{\partial}{\partial \tilde{\nu}(y)} \frac{1}{|\Psi_i(x) \pm \tau\nu(\Psi_i(x)) - \Psi_j(y)|} d\lambda^2(y)$$

$$- \sum_{q=1}^{2} \int_{B_1^{\mathbb{R}^2}(0)} \partial_1^{t_1} \partial_2^{t_2} \left(F(\Psi_j(y))\right) G_2(y) H_2(x) \frac{\tilde{\nu}_q(y)}{\tilde{\nu}_3(y)} \frac{\partial}{\partial y_q} \frac{1}{|\Psi_i(x) \pm \tau\nu(\Psi_i(x)) - \Psi_j(y)|} d\lambda^2(y).$$

So using again integration by parts for $q = 1, 2$, we finally have to estimate the $L^2(B_1^{\mathbb{R}^2}(0))$-norm of the following two types of terms

$$H(x) \int_{B_1^{\mathbb{R}^2}(0)} \partial_1^{r_1} \partial_2^{r_2} \left(F(\Psi_j(y))\right) G(y) \frac{1}{|\Psi_i(x) \pm \tau\nu(\Psi_i(x)) - \Psi_j(y)|} d\lambda^2(y), \qquad (4.38)$$

$$H(x) \int_{B_1^{\mathbb{R}^2}(0)} \partial_1^{r_1} \partial_2^{r_2} \left(F(\Psi_j(y))\right) G(y) \frac{\partial}{\partial \tilde{\nu}(y)} \frac{1}{|\Psi_i(x) \pm \tau\nu(\Psi_i(x)) - \Psi_j(y)|} d\lambda^2(y), \qquad (4.39)$$

with $0 \leq |r| \leq m+1$, $G \in C^0\left(\overline{B_1^{\mathbb{R}^2}(0)}\right)$ and $H \in C^2\left(\overline{B_1^{\mathbb{R}^2}(0)}\right)$. So in order to estimate the terms in (4.19) we finally have to estimate the $L^2(B_1^{\mathbb{R}^2}(0))$-norm of the terms (4.38) and (4.39). We have

$$\left\| H(\,\cdot\,) \int_{B_1^{\mathbb{R}^2}(0)} \partial_1^{r_1} \partial_2^{r_2} \left(F(\Psi_j(y))\right) G(y) \frac{1}{|\Psi_i(\,\cdot\,) \pm \tau\nu(\Psi_i(\,\cdot\,)) - \Psi_j(y)|} d\lambda^2(y) \right\|_{L^2(B_1^{\mathbb{R}^2}(0))}$$

$$\leq \|H(\,\cdot\,)\|_{C^0(\overline{B_1^{\mathbb{R}^2}(0)})} \left\| \int_{B_1^{\mathbb{R}^2}(0)} \partial_1^{r_1} \partial_2^{r_2} \left(F(\Psi_j(y))\right) G(y) \frac{1}{|\Psi_i(\,\cdot\,) \pm \tau\nu(\Psi_i(\,\cdot\,)) - \Psi_j(y)|} d\lambda^2(y) \right\|_{L^2(B_1^{\mathbb{R}^2}(0))}$$

$$= \|H\|_{C^0(\overline{B_1^{\mathbb{R}^2}(0)})} \left\| U_1[\chi_{U_j} \frac{\partial_1^{r_1} \partial_2^{r_2} \left(F(\Psi_j)\right) G}{J_j} \circ \Psi_j^{-1}](\Psi_i(\,\cdot\,) \pm \tau\nu(\Psi_i(\,\cdot\,))) \right\|_{L^2(B_1^{\mathbb{R}^2}(0))}$$

$$\leq \left\| U_1[\chi_{U_j} \frac{\partial_1^{r_1} \partial_2^{r_2} \left(F(\Psi_j)\right) G}{J_j} \circ \Psi_j^{-1}](\,\cdot\, \pm \tau\nu(\,\cdot\,)) \right\|_{L^2(\partial \Sigma)}$$

$$\leq C_9 \left\| \chi_{U_j} \frac{\partial_1^{r_1} \partial_2^{r_2} \left(F(\Psi_j)\right) G}{J_j} \circ \Psi_j^{-1} \right\|_{L^2(\partial \Sigma)} \leq C_9 \frac{\|G\|_{C^0(\overline{B_1^{\mathbb{R}^2}(0)})}}{c_1^j} \|F\|_{H^{m+1,2}(\partial \Sigma)},$$

where we used the equivalent norm on $L^2(\partial \Sigma)$ introduced in Lemma 2.2.6 and the constants from Lemma 4.5.2 and Lemma 2.1.4. Similar we estimate

$$\left\| H(\,\cdot\,) \int_{B_1^{\mathbb{R}^2}(0)} \partial_1^{r_1} \partial_2^{r_2} \left(F(\Psi_j(y))\right) G(y) \frac{\partial}{\partial \tilde{\nu}(y)} \frac{1}{|\Psi_i(\,\cdot\,) \pm \tau\nu(\Psi_i(\,\cdot\,)) - \Psi_j(y)|} d\lambda^2(y) \right\|_{L^2(B_1^{\mathbb{R}^2}(0))}$$

$$\leq \|H(\,\cdot\,)\|_{C^0(\overline{B_1^{\mathbb{R}^2}(0)})}$$

$$\cdot \left\| \int_{B_1^{\mathbb{R}^2}(0)} \partial_1^{r_1} \partial_2^{r_2} \left(F(\Psi_j(y)) \right) G(y) \frac{\partial}{\partial \tilde{\nu}(y)} \frac{1}{|\Psi_i(\,\cdot\,) \pm \tau\nu(\Psi_i(\,\cdot\,)) - \Psi_j(y)|} d\lambda^2(y) \right\|_{L^2(B_1^{\mathbb{R}^2}(0))}$$

$$= \|H\|_{C^0(\overline{B_1^{\mathbb{R}^2}(0)})} \left\| \frac{\partial U_1}{\partial \nu} [\chi_{U_j} \frac{\partial_1^{r_1} \partial_2^{r_2} (F(\Psi_j))) G}{J_j} \circ \Psi_j^{-1}](\Psi_i(\,\cdot\,) \pm \tau\nu(\Psi_i(\,\cdot\,))) \right\|_{L^2(B_1^{\mathbb{R}^2}(0))}$$

$$\leq \left\| \frac{\partial U_1}{\partial \nu} [\chi_{U_j} \frac{\partial_1^{r_1} \partial_2^{r_2} (F(\Psi_j))) G}{J_j} \circ \Psi_j^{-1}](\,\cdot\, \pm \tau\nu(\,\cdot\,)) \right\|_{L^2(\partial\Sigma)}$$

$$\leq C_{10} \left\| \chi_{U_j} \frac{\partial_1^{r_1} \partial_2^{r_2} (F(\Psi_j))) G}{J_j} \circ \Psi_j^{-1} \right\|_{L^2(\partial\Sigma)} \leq C_{10} \frac{\|G\|_{C^0(\overline{B_1^{\mathbb{R}^2}(0)})}}{c_1^j} \|F\|_{H^{m+1,2}(\partial\Sigma)}.$$

So recalling formula (4.19) we find C_{15} of the form

$$C_{15} := \sum_{i,j=1}^{N} \sum_{s_1+s_2=0}^{m} \sum_{k_1=0}^{s_1} \sum_{k_2=0}^{s_2} \binom{s_1}{k_1} \binom{s_2}{k_2} \|w_i\|_{C^m(\partial\Sigma)} (C_9 + C_{10}) \left\| \frac{G_j}{J_j} \right\|_{C^0(\overline{B_1^{\mathbb{R}^2}(0)})} \|H_i\|_{C^0(\overline{B_1^{\mathbb{R}^2}(0)})}$$

where G_j is a sum of products from the partial derivatives of Ψ_j, $\tilde{\nu}$ and J_j up to order $m+1$ which can be obtained from the considerations above. Similar H_i consist of partial derivatives of Ψ_i and ν up to order $m-1$. For $|s|=0$ we can use Lemma 4.5.1 and so the proof for $L_3^{\pm\tau}$ is done. The argumentation above translates to the case of $L_1^{\pm\tau}$. It even simplifies because we do not have to get rid of the normal derivative at the beginning of the proof. More precisely we get the terms as in (4.19) without $\frac{\partial}{\partial \tilde{\nu}(y)}$ and thus we gain one iteration and consequently one order of differentiation, i.e. we can proof the convergence for $F \in H^{m,2}(\partial\Sigma)$ and an outer $C^{m+1,\alpha}$-domain Σ. We need at least an outer $C^{m+1,\alpha}$-domain instead of an outer C^{m+1}-domain in order to ensure the convergence in $C^m(\partial\Sigma)$-norm, which is used in the proof. It is left to investigate $L_2^{\pm\tau}$ and $L_4^{\pm\tau}$. In order to apply the argumentation from $L_1^{\pm\tau}$ to $L_2^{\pm\tau}$ and from $L_3^{\pm\tau}$ to $L_4^{\pm\tau}$ we have to make the following consideration to get rid of the normal derivative before we can apply the techniques used in the previous part of the proof. We will illustrate this procedure at

$$\lim_{\tau \to 0^+} \|L_4^{\pm\tau}[F]\|_{L^2(\partial\Sigma)} = 0,$$

for $F \in H^{m+2,2}(\partial\Sigma)$ and $\partial\Sigma$ a C^{m+3}-surface. We have to prove

$$\left\| \frac{\partial U_2}{\partial \nu}(\,\cdot\, \pm \tau\nu(\,\cdot\,)) \right\|_{L^2(\partial\Sigma)} \leq C_{16} \|F\|_{H^{m+2,2}(\partial\Sigma)},$$

for all $\tau \in (0, \tau_0]$ with $0 < C_{16} < \infty$ independent of τ. Then the argumentation from the beginning of this proof applies and we are done. We have for $x \in \partial\Sigma$, $\tau \in (0, \tau_0]$ and $F \in C^{m+2}(\partial\Sigma)$

$$\frac{\partial U_2}{\partial \nu}(x \pm \tau\nu(x)) = \langle \nu(x), (\nabla U_2)(x \pm \tau\nu(x)) \rangle = \langle \nu(x \pm \tau\nu(x)), (\nabla U_2)(x \pm \tau\nu(x)) \rangle$$

$$=\langle \nu(z), \nabla U_2(z)\rangle\Big|_{z=x\pm\tau\nu(x)} = \langle \nu(z), \nabla \int_{\partial\Sigma} F(y)\frac{\partial}{\partial\nu(y)}\frac{1}{|z-y|}dH^2(y)\rangle\Big|_{z=x\pm\tau\nu(x)}.$$

Because $\tau \neq 0$ we can interchange integration and differentiation, see Lemma 2.4.8, and we get

$$\frac{\partial U_2}{\partial \nu}(x \pm \tau\nu(x)) = \int_{\partial\Sigma} F(y)\frac{\partial}{\partial\nu(y)}\frac{\partial}{\partial\nu(z)}\frac{1}{|z-y|}dH^2(y)\Big|_{z=x\pm\tau\nu(x)}$$

$$= -\int_{\partial\Sigma} F(y)\frac{\partial}{\partial\nu(y)}\frac{\langle z-y, \nu(z)\rangle}{|z-y|^3}dH^2(y)\Big|_{z=x\pm\tau\nu(x)}$$

$$= -\sum_{i=1}^{N}\int_{B_1^{\mathbb{R}^2}(0)} w_i(\Psi_i(y'))F(\Psi_i(y'))$$

$$\frac{\partial}{\partial\tilde\nu(y')}\frac{\langle x\pm\tau\nu(x) - \Psi_i(y'), \nu(x\pm\tau\nu(x))\rangle}{|x\pm\tau\nu(x) - \Psi_i(y')|^3}d\lambda^2(y')$$

$$= -\sum_{i=1}^{N}\int_{B_1^{\mathbb{R}^2}(0)} w_i(\Psi_i(y'))F(\Psi_i(y'))\frac{\partial}{\partial\tilde\nu(y')}\frac{\langle x\pm\tau\nu(x) - \Psi_i(y'), \nu(x)\rangle}{|x\pm\tau\nu(x) - \Psi_i(y')|^3}d\lambda^2(y')$$

$$= -\sum_{i=1}^{N}\sum_{j=1}^{3}\int_{B_1^{\mathbb{R}^2}(0)} w_i(\Psi_i(y'))F(\Psi_i(y'))\frac{\partial}{\partial\tilde\nu(y')}\left(b_j(y')c_j(x)\frac{\partial}{\partial y'_j}\frac{1}{|x\pm\tau\nu(x) - \Psi_i(y')|}\right)d\lambda^2(y')$$

where b_j and c_j for $j = 1, 2, 3$ and $i \in \{1, \ldots, N\}$, are defined by

$$(bc)(x, y') = (D\Psi_i(y'))^{-1}\nu(x),$$

where $(bc)(x, y') := (b_1(x)c_1(y'), b_2(x)c_2(y'), b_3(x)c_3(y'))^T$. Here the c_j are even independent from $\tau \in (0, \tau_0]$. Now we use integration by parts and the argumentation for $L_3^{\pm\tau}$ applies with one additional order of differentiability for F and Σ. In the same way $L_2^{\pm\tau}$ can be reduced to $L_1^{\pm\tau}$, also with one additional order of differentiability for F and Σ. Now the proof is complete. □

Before we come to the applications in the next section, we want to state the following remark.

Remark 4.6.2. Due to the techniques used in the previous proof, we can see why a higher regularity of F and Σ leads to a higher regularity of the corresponding potentials and their convergence. The reason is that the integral kernel $\frac{1}{|x-y|}$ allows us to translate differentiation with respect to x into differentiation with respect to y. Then we can put the differential operators to F and Ψ_i with help of integration by parts.

4.7 Application of the Limit Formulae to Geomathematics

In this section we present an application of the limit formulae and jump relations to geomathematics. The double and single layer potential as well as their limit formulae are important in geomathematics, because they are used in order to generate wavelets. In [FM03], such a wavelet approach to C^2 surfaces is presented, using the convergence in $L^2(\partial\Sigma)$. Our goal is to prove density of certain function systems from geomathematics in the spaces $H^{m,2}(\partial\Sigma)$. The main result is a direct consequence of the results proved in this chapter and a generalization of the results contained in [FK80], [FM03] and [FM04]. Here the authors prove that the function system of mass point representations and the function system of outer harmonics are dense in $L^2(\partial\Sigma)$. The result is used in order to approximate solutions to the Dirichlet problem for the Laplace equation, i.e., the homogeneous Poisson equation, in Σ or D, respectively. We prove that the function system of mass point representations as well as the function systems of inner and outer harmonics are a dense subspace of $H^{m,2}(\partial\Sigma)$ for arbitrary $m \in \mathbb{N}$. For example, the results proved in this section could lead to a faster convergence of those approximations. At first we state the result from [FM03] and [FM04], which we will extend in this section.

Lemma 4.7.1. Let Σ be an outer C^2-domain and $(x_k)_{k\in\mathbb{N}}$ a fundamental system in Σ or D. Then the system of mass point representations

$$\left(\left.\frac{1}{|x_k - \cdot|}\right|_{\partial\Sigma}\right)_{k\in\mathbb{N}},$$

defined by Definition 2.6.1, is dense in $L^2(\partial\Sigma)$. Furthermore the system of outer harmonics

$$\left(\left.H^\alpha_{-n-1,k}\right|_{\partial\Sigma}\right)_{n=0,1,\ldots;k=1,\ldots,2n+1},$$

defined in Definition 2.6.2, is dense in $L^2(\partial\Sigma)$.

Proof. For this result see e.g. [FM04]. □

In this section we want to extend this result to the case of $H^{m,2}(\partial\Sigma)$, $m \in \mathbb{N}$, $m \geq 1$. Moreover, we prove the result also for the inner harmonics for $H^{m,2}(\partial\Sigma)$, $m \in \mathbb{N}$. Before we can do this, we need to prove several lemmata. At first we prove a result for the Hilbert space adjoints $\left(U_1^{\pm\tau}\right)^*$ of U_1 with respect to the $L^2(\partial\Sigma)$ scalar product, defined in Lemma 4.5.2. It

4.7. APPLICATION OF THE LIMIT FORMULAE TO GEOMATHEMATICS

states that the $H^{m,2}(\partial\Sigma)$-norm of $\left(U_1^{\pm\tau}\right)^*[F]$ can be estimated by the $H^{m,2}(\partial\Sigma)$-norm of F with a constant independent of $\tau \in [0, \tau_0]$.

Lemma 4.7.2. Let Σ be an outer C^{m+2} domain, $m \in \mathbb{N}$, $m \geq 1$. Then for each $\tau \in [0, \tau_0]$ we have $\left(U_1^{\pm\tau}\right)^* : H^{m,2}(\partial\Sigma) \to H^{m,2}(\partial\Sigma)$ with

$$\| \left(U_1^{\pm\tau}\right)^*[F]\|_{H^{m,2}(\partial\Sigma)} \leq C_{17}\|F\|_{H^{m,2}(\partial\Sigma)},$$

for all $F \in H^{m,2}(\partial\Sigma)$ and $\tau \in [0, \tau_0]$, where $0 < C_{17} < \infty$ is a constant independent of $\tau \in [0, \tau_0]$.

Proof. Recall the Definition of the adjoint operators from Lemma 4.5.2. It is clear that the operators are linear. The desired estimate can be proved analogous to those from Theorem 4.6.1. Therefore, we just want to mention the crucial points. Recall the proof of Theorem 4.6.1. Corresponding to (4.19), we now have to estimate the $L^2(\partial\Sigma)$-norm of terms of the form

$$\partial_1^{s_1}\partial_2^{s_2} \int_{B_1^{\mathbb{R}^2}(0)} w_i(\Psi_i(y))F(\Psi_i(y))\frac{1}{|\Psi_i(y) \pm \tau\nu(\Psi_i(y)) - \Psi_j(x)|}J_i(y)d\lambda^2(y),$$

for $F \in C^m(\partial\Sigma)$, $0 \leq s_1 + s_2 \leq m$, by the $H^{m,2}(\partial\Sigma)$-norm of F times a constant $0 < C_{17} < \infty$ independent of $\tau \in (0, \tau_0]$. Recall that we have

$$\nu(\Psi_i) = \frac{\partial_1\Psi_i \times \partial_2\Psi_i}{|\partial_1\Psi_i \times \partial_2\Psi_i|},$$

for all $i = 1, \ldots, N$, see also Lemma 2.1.6, and we obtain the transformed terms

$$\partial_1^{s_1}\partial_2^{s_2} \int_{B_1^{\mathbb{R}^2}(0)} w_i(\Psi_i(y))F(\Psi_i(y))\frac{1}{|\Psi^{\pm\tau}(y) - \Psi_j(x)|}J_i(y)d\lambda^2(y),$$

for $F \in C^m(\partial\Sigma)$, $0 \leq s_1 + s_2 \leq m$ and $\tau \in (0, \tau_0]$, if we set

$$\Psi^{\pm\tau}(y) := \Psi_i(y) \pm \tau\frac{\partial_1\Psi_i(y) \times \partial_2\Psi_i(y)}{|\partial_1\Psi_i(y \times \partial_2\Psi_i(y)|}.$$

for all $y \in B_1^{\mathbb{R}^3}(0)$. We have $\Psi^{\pm\tau} \in C^{m+1}\left(\overline{B_1^{\mathbb{R}^3}(0)}\right)$ with $\mathrm{Det}\left(D\Psi\right) > 0$ on $\overline{B_1^{\mathbb{R}^3}(0)}$ for $\tau \in [0, \tau_0']$ if $0 < \tau_0' < \tau_0$ is small enough. Consequently

$$\left\{\partial_1\Psi^{\pm\tau}(y), \partial_2\Psi^{\pm\tau}(y), \partial_3\Psi^{\pm\tau}(y)\right\}$$

forms a basis of $\mathbb{R}^3$ for each $y \in B_1^{\mathbb{R}^2}(0)$ and $\tau \in (0, \tau_0']$, which is $C^m(B_1^{\mathbb{R}^2}(0))$ for the outer C^{m+2}-domain Σ and bounded in $C^m(B_1^{\mathbb{R}^2}(0))$ norm independent of τ. So we can translate differentiation with respect to x to differentiation with respect to y by solving the linear system of equations

$$a^k(x,y) D\Psi^{\pm\tau}(y) = \partial_k \Psi_i(x)$$

and get a solution with coefficient functions in $C^m\left(\overline{B_1^{\mathbb{R}^2}(0)}\right)$. For details see the proof of Theorem 4.6.1. Now we apply integration by parts and apply the reduction method for the higher order derivatives in direction y_3. Note that we can do this because $\Psi^{\pm\tau} \in C^2(B_1^{\mathbb{R}^3}(0))$ and the differential geometric functions g, g^{pq} and g_{pq} are elements of $C^m(B_1^{\mathbb{R}^3}(0))$, $1 \leq p,q \leq 3$, and also bounded in $C^m(B_1^{\mathbb{R}^2}(0))$ norm independent of τ by their definition. Furthermore, $g^{33} \neq 0$ on $\overline{B_1^{\mathbb{R}^2}(0)}$ for $0 < \tau_0' < \tau_0$ small enough. So all techniques can be applied and we end up with terms of the form

$$H(x) \int_{B_1^{\mathbb{R}^2}(0)} \partial_1^{r_1} \partial_2^{r_2} \left(F(\Psi_j(y))\right) G(y) \frac{1}{|\Psi_i(y) \pm \tau\nu(\Psi_i(y)) - \Psi_j(x)|} d\lambda^2(y), \tag{4.40}$$

$$H(x) \int_{B_1^{\mathbb{R}^2}(0)} \partial_1^{r_1} \partial_2^{r_2} \left(F(\Psi_j(y))\right) G(y) \frac{\partial}{\partial \tilde{\nu}(y)} \frac{1}{|\Psi_i(y) \pm \tau\nu(\Psi_i(y)) - \Psi_j(x)|} d\lambda^2(y), \tag{4.41}$$

where $0 \leq |r| \leq m$, $G \in C^0\left(\overline{B_1^{\mathbb{R}^2}(0)}\right)$ and $H \in C^1\left(\overline{B_1^{\mathbb{R}^2}(0)}\right)$. The vectorfield $\tilde{\nu}$ is defined analogous as in Theorem 4.6.1 by

$$\tilde{\nu} := \left(D\Psi^{\pm\tau}\right)^{-1} \nu(\Psi_i),$$

on $B_1^{\mathbb{R}^2}(0)$. So we have $\tilde{\nu} \in C^m\left(\overline{B_1^{\mathbb{R}^2}(0)}\right)$, with bounded norm independent of τ. With the reasoning from the proof of Theorem 4.6.1, we can transform $\frac{\partial}{\partial y_3}$ to $\frac{\partial}{\partial \tilde{\nu}(y)}$, because $D\Psi^{\pm\tau}$ has full range. The terms (4.40) and (4.41) correspond to (4.38) and (4.39) in the proof of Theorem 4.6.1 and can be estimated similar with help of the results about $\left(U_1^{\pm\tau}\right)^*$ and $\left(\frac{\partial U_1}{\partial \nu}^{\pm\tau}\right)^*$ from Lemma 4.5.2. For $|s| = 0$ we refer directly to Lemma 4.5.2 and we are done. Recall that $(U_1^0)^* = U_1$ is already treated in Theorem 4.6.1. $\square$

Note that $U_1[F]$ is self adjoint. Now we extend U_1 to an operator onto $(H^{m,2}(\partial\Sigma))'$. This operator is then defined as the Banach space adjoint of $\left(U_1^{\pm\tau}\right)^*$ in the Banach space $(H^{m,2}(\partial\Sigma))'$.

Lemma 4.7.3. Let Σ be an outer C^{m+2} domain, $m \in \mathbb{N}$, $m \geq 1$. Then we define for each $\tau \in [0, \tau_0]$ a linear and bounded operator

$$U_1^{\pm\tau} : \left(H^{m,2}(\partial\Sigma)\right)' \to \left(H^{m,2}(\partial\Sigma)\right)',$$

by

$$U_1^{\pm\tau}[F](G) := F\left(\left(U_1^{\pm\tau}\right)^*[G]\right), \tag{4.42}$$

4.7. APPLICATION OF THE LIMIT FORMULAE TO GEOMATHEMATICS

for all $G \in H^{m,2}(\partial\Sigma)$. We have that

$$\|U_1^{\pm\tau}[F]\|_{(H^{m,2}(\partial\Sigma))'} \leq C_{18}\|F\|_{(H^{m,2}(\partial\Sigma))'},$$

for all $F \in (H^{m,2}(\partial\Sigma))'$ and $\tau \in [0, \tau_0]$, where C_{18} and $(U_1^{\pm\tau})^*$ are taken from Lemma 4.7.2.

Proof. Due to Lemma 4.7.2 we have that $U_1^{\pm\tau}$ is well defined and we can estimate

$$|U_1^{\pm\tau}[F](G)| = \left|F\left((U_1^{\pm\tau})^*[G]\right)\right| \leq \|F\|_{(H^{m,2}(\partial\Sigma))'}\|(U_1^{\pm\tau})^*[G]\|_{H^{m,2}(\partial\Sigma)}$$
$$\leq C_{19}\|F\|_{(H^{m,2}(\partial\Sigma))'}\|G\|_{H^{m,2}(\partial\Sigma)},$$

for all $F \in (H^{m,2}(\partial\Sigma))'$ and $G \in H^{m,2}(\partial\Sigma)$. Thus the norm estimate holds. $\square$

Using the results from Section 4.6, we are able to prove that the limit formula of U_1 even holds for $F \in (H^{m,2}(\partial\Sigma))'$.

Theorem 4.7.4. Let Σ be an outer C^{m+2} domain, $m \in \mathbb{N}$, $m \geq 1$, and $F \in (H^{m,2}(\partial\Sigma))'$. Then we have

$$\lim_{\tau \to 0+} \|U_1^{\pm\tau}[F] - U_1^0[F]\|_{(H^{m,2}(\partial\Sigma))'} = 0. \tag{4.43}$$

Proof. Let $F \in (H^{m,2}(\partial\Sigma))'$. Because $L^2(\partial\Sigma)$ is dense in $(H^{m,2}(\partial\Sigma))'$, we can choose a sequence $(F_n)_{n\in\mathbb{N}} \subset L^2(\partial\Sigma)$ with $F_n \to F$ in $(H^{m,2}(\partial\Sigma))'$. We now estimate with help of Lemma 4.7.2

$$\|U_1^{\pm\tau}[F] - U_1^0[F]\|_{(H^{m,2}(\partial\Sigma))'}$$
$$\leq \|U_1^{\pm\tau}[F] - U_1^{\pm\tau}[F_n]\|_{(H^{m,2}(\partial\Sigma))'} + \|U_1^{\pm\tau}[F_n] - U_1^0[F_n]\|_{(H^{m,2}(\partial\Sigma))'}$$
$$+ \|U_1^0[F_n] - U_1^0[F]\|_{(H^{m,2}(\partial\Sigma))'}$$
$$\leq 2C_{18}\|F_n - F\|_{(H^{m,2}(\partial\Sigma))'} + \|U_1^{\pm\tau}[F_n] - U_1^0[F_n]\|_{(H^{m,2}(\partial\Sigma))'}.$$

The first term converges to zero as n tends to infinity. For the second we have

$$\|U_1^{\pm\tau}[F_n] - U_1^0[F_n]\|_{(H^{m,2}(\partial\Sigma))'}$$
$$= \sup_{G \in H^{m,2}(\partial\Sigma)} \frac{|U_1^{\pm\tau}[F_n](G) - U_1^0[F_n](G)|}{\|G\|_{H^{m,2}(\partial\Sigma)}}$$
$$= \sup_{G \in H^{m,2}(\partial\Sigma)} \frac{|F_n((U_1^{\pm\tau})^*[G]) - F_n((U_1^0)^*[G])|}{\|G\|_{H^{m,2}(\partial\Sigma)}}$$

$$=\sup\nolimits_{G\in H^{m,2}(\partial\Sigma)} \frac{|\int_{\partial\Sigma} F_n(x)\cdot (U_1^{\pm\tau})^*[G](x)\,dH^2(x) - \int_{\partial\Sigma} F_n(x)\cdot (U_1^0)^*[G](x)\,dH^2(x)|}{\|G\|_{H^{m,2}(\partial\Sigma)}}$$

$$=\sup\nolimits_{G\in H^{m,2}(\partial\Sigma)} \frac{|\int_{\partial\Sigma} U_1[F_n](x\pm\nu(x))\cdot G(x)\,dH^2(x) - \int_{\partial\Sigma} U_1[F_n](x)\cdot G(x)\,dH^2(x)|}{\|G\|_{H^{m,2}(\partial\Sigma)}}$$

$$\leq \sup\nolimits_{G\in H^{m,2}(\partial\Sigma)} \frac{\|U_1[F_n](\,\cdot\,\pm\nu(\,\cdot\,)) - U_1[F_n](\,\cdot\,)\|_{L^2(\partial\Sigma)}\|G\|_{L^2(\partial\Sigma)}}{\|G\|_{H^{m,2}(\partial\Sigma)}}$$

$$\leq \|U_1[F_n](\,\cdot\,\pm\nu(\,\cdot\,)) - U_1[F_n](\,\cdot\,)\|_{L^2(\partial\Sigma)}$$

which converges to zero if τ does, by Lemma 4.5.1. Here we used that we can replace the dual paring by the $L^2(\partial\Sigma)$ scalar product, i.e., the integral over $\partial\Sigma$, for $F_n \in L^2(\partial\Sigma)$. Now the proof is done. □

We now define for $m \in \mathbb{N}$, $m \geq 1$, a differential operator $D^{2m} : H^{m,2}(\partial\Sigma) \to (H^{m,2}(\partial\Sigma))'$.

Lemma 4.7.5. Let $m \in \mathbb{N}$, $m \geq 1$, and Σ be a $C^{m,1}$-surface. Then we define the linear and bounded operator $D^{2m} : H^{m,2}(\partial\Sigma) \to (H^{m,2}(\partial\Sigma))'$ by

$$D^{2m}[F](G) := \langle F, G \rangle_{H^{m,2}(\partial\Sigma)},$$

for all $G \in H^{m,2}(\partial\Sigma)$.

Proof. Obviously we have

$$|D^{2m}[F](G)| = |\langle F, G\rangle_{H^{m,2}(\partial\Sigma)}| \leq \|F\|_{H^{m,2}(\partial\Sigma)}\|G\|_{H^{m,2}(\partial\Sigma)}, \qquad (4.44)$$

for all $F, G \in H^{m,2}(\partial\Sigma)$ and the proof is done. □

Having in mind the Gelfand triple

$$H^{m,2}(\partial\Sigma) \subset L^2(\partial\Sigma) \subset \left(H^{m,2}(\partial\Sigma)\right)',$$

the operator D^{2m} is the standard continuous embedding of $H^{m,2}(\partial\Sigma)$ into $(H^{m,2}(\partial\Sigma))'$. Moreover, we have the following.

Lemma 4.7.6. Let $m \in \mathbb{N}$, $m \geq 1$, and Σ be a $C^{2m-1,1}$-surface. Then for each $F \in C^\infty(\partial\Sigma)$ there exists a function $D^{2m}F \in L^2(\partial\Sigma)$ such that

$$D^{2m}[F](G) = \langle D^{2m}F, G\rangle_{L^2(\partial\Sigma)},$$

for all $G \in H^{m,2}(\partial\Sigma)$, where D^{2m} is the operator, defined in Lemma 4.7.5.

4.7. APPLICATION OF THE LIMIT FORMULAE TO GEOMATHEMATICS

Proof. We have by the definition of D^{2m}, see Lemma 4.7.5, that

$$D^{2m}[F](G) = \langle F, G \rangle_{H^{m,2}(\partial \Sigma)} = \sum_{i=1}^{N} \langle (w_i F)(\Psi_i(y,0)), (w_i G)(\Psi_i(y,0)) \rangle_{H^{m,2}(B_1^{\mathbb{R}^2}(0))}$$

$$= \sum_{i=1}^{N} \sum_{|s|=0}^{m} \int_{B_1^{\mathbb{R}^2}(0)} \partial^s \Big((w_i F)(\Psi_i(y,0)) \Big) \cdot \partial^s \Big((w_i G)(\Psi_i(y,0)) \Big) d\lambda^2(y)$$

$$= -\sum_{i=1}^{N} \sum_{|s|=0}^{m} \int_{B_1^{\mathbb{R}^2}(0)} \partial^{2s} \Big((w_i F)(\Psi_i(y,0)) \Big) \cdot (w_i G)(\Psi_i(y,0)) \, d\lambda^2(y)$$

$$= -\sum_{i=1}^{N} \sum_{|s|=0}^{m} \int_{B_1^{\mathbb{R}^2}(0)} \partial^{2s} \Big((w_i F)(\Psi_i(y,0)) \Big) \cdot \frac{w_i(\Psi_i(y,0))}{J_i(y)} \cdot G(\Psi_i(y,0)) J_i(y) \, d\lambda^2(y)$$

$$= -\sum_{i=1}^{N} \sum_{|s|=0}^{m} \int_{\partial \Sigma} \Big(\partial^{2s}((w_i F) \circ \Psi_i) \Big)(\Psi_i^{-1}(x)) \cdot \frac{w_i(x)}{J_i(\Psi_i^{-1}(x))} \cdot G(x) \, dH^2(x)$$

$$= \int_{\partial \Sigma} \underbrace{\left(-\sum_{i=1}^{N} \chi_{U_i \cap \partial \Sigma} \cdot \sum_{|s|=0}^{m} \Big(\partial^{2s}((w_i F) \circ \Psi_i) \Big)(\Psi_i^{-1}(x)) \cdot \frac{w_i(x)}{J_i(\Psi_i^{-1}(x))} \right)}_{=: D^{2m} F} \cdot G(x) \, dH^2(x),$$

which is a function in $L^2(\partial \Sigma)$, because $F \in C^{\infty}(\partial \Sigma)$, $\Psi_i \in H^{2m,\infty}(B_1^{\mathbb{R}^2}(0))$ see Lemma 2.2.5, $w_i \in C^{\infty}(U_i)$ and $0 < c_1^i < J_i < c_2^i < \infty$, see Lemma 2.1.4. □

Consequently, D^{2m} is the extension of the differential operator defined above to a space of generalized functions. We need one final lemma about the potential of the single layer.

Lemma 4.7.7. Let $\partial \Sigma$ be a $C^{m+1,\alpha}$-surface, $m \in \mathbb{N}$, $m \geq 1$, $0 < \gamma < \beta \leq \alpha \leq 1$. Then for each $F \in C^{m,\beta}(\partial \Sigma)$ there exists a $G \in C^{m-1,\gamma}(\partial \Sigma)$ such that

$$F = U_1[G],$$

on $\partial \Sigma$.

Proof. Let $F \in C^{m,\beta}(\partial \Sigma)$ be given. Then the unique solution U of the Dirichlet problem in D and Σ with boundary value F has continuous normal derivatives on $\partial \Sigma$, see [CK83, Theorem 3.29], because $F \in C^{m,\alpha}(\partial \Sigma) \subset C^{1,\alpha}(\partial \Sigma)$. We define

$$G(x) := \frac{\partial U^+}{\partial \nu} - \frac{\partial U^-}{\partial \nu},$$

for all $x \in \partial\Sigma$, where $\frac{\partial U^{\pm}}{\partial \nu}$ denotes the normal derivatives on $\partial\Sigma$ of the solution in Σ, denoted by U^+, or of the solution in D, denoted by U^-. We know that U is given by the single layer potential $U_1[G]$ on $\mathbb{R}^3$, see [CK83, Theorem 3.30]. So $U_1[G] = F$ on $\partial\Sigma$ and it is left to show that $G \in C^{m-1,\gamma}(\partial\Sigma)$ for some $0 < \gamma < \beta$. Therefore we need to prove that $\frac{\partial U^{\pm}}{\partial \nu} \in C^{m-1,\gamma}(\partial\Sigma)$. We know that $U_2[F]$ is a harmonic function in $C^{\infty}(\mathbb{R}^3 \setminus \partial\Sigma)$ and so it solves the Neumann problem in D and Σ with boundary value $\frac{\partial U_2}{\partial \nu}[F] \in C^{m-1,\delta}(\partial\Sigma)$ for all $0 < \delta < \beta$, see Theorem 4.4.1. As shown in [CK83, Theorem 3.16 and Theorem 3.25] this solution can be represented as a single layer potential $U_1[H_1]$ in D and $U_1[H_2]$ in Σ. Moreover, [Gün57, p. 170] gives $H_1, H_2 \in C^{m-1,\gamma}(\partial\Sigma)$ for all $0 < \gamma < \delta < \beta$. We define two functions V and W on $\mathbb{R}^3 \setminus \partial\Sigma$ by

$$V(x) := \begin{cases} U_1[H_1](x) - U_2[F](x) + 4\pi U(x), & x \in \Sigma, \\ U_1[H_1](x) - U_2[F](x), & x \in D, \end{cases}$$

and

$$W(x) := \begin{cases} U_1[H_2](x) - U_2[F](x), & x \in \Sigma, \\ U_1[H_2](x) - U_2[F](x) + 4\pi U(x), & x \in D. \end{cases}$$

We have

$$\lim_{\tau \to 0^+} V(x + \tau\nu(x)) - V(x - \tau\nu(x)) = 0,$$

$$\lim_{\tau \to 0^+} W(x + \tau\nu(x)) - W(x - \tau\nu(x)) = 0,$$

for all $x \in \partial\Sigma$, see Theorem 4.2.2. Furthermore we have $V = 0$ on D and $W = 0$ on Σ. Together this yields V as the unique solution of the homogeneous Dirichlet problem in Σ and W as the unique solution of the homogeneous Dirichlet problem in D. Consequently V and W are zero on all of $\mathbb{R}^3$. Using the limit formulae we get

$$0 = \lim_{\tau \to 0^+} \frac{\partial V}{\partial \nu}(x + \tau\nu(x)) - \frac{\partial U_1}{\partial \nu}[H_1](x) + \frac{\partial U_2}{\partial \nu}[F](x) - 4\pi \frac{\partial U^+}{\partial \nu}(x)$$

$$= \lim_{\tau \to 0^+} \frac{\partial U_1}{\partial \nu}[H_1](x) - \frac{\partial U_2}{\partial \nu}[F](x) + 4\pi \frac{\partial U^+}{\partial \nu}(x)$$

$$= \frac{\partial U_1}{\partial \nu}[H_1](x) - \frac{\partial U_2}{\partial \nu}[F](x) + 4\pi \frac{\partial U^+}{\partial \nu}(x),$$

for all $x \in \partial\Sigma$ and analogously

$$\frac{\partial U_1}{\partial \nu}[H_2](x) - \frac{\partial U_2}{\partial \nu}[F](x) + 4\pi \frac{\partial U^-}{\partial \nu} = 0,$$

4.7. APPLICATION OF THE LIMIT FORMULAE TO GEOMATHEMATICS

for all $x \in \partial\Sigma$. Finally we have proved that $\frac{\partial U}{\partial \nu}^\pm$ equals an $C^{m-1,\gamma}(\partial\Sigma)$ function, because $\frac{\partial U_1}{\partial \nu}[H_1]$, $\frac{\partial U_1}{\partial \nu}[H_2]$ and $\frac{\partial U_2}{\partial \nu}[F]$ are elements of $C^{m-1,\gamma}(\partial\Sigma)$, see Theorem 4.4.1, and the proof is done. □

Moreover, the function G from the previous lemma is uniquely determined, see [CK83, Theorem 3.30]. We finally extend Lemma 4.7.1 in the main result of this section.

Theorem 4.7.8. *Let Σ be an outer C^2-domain for $m = 0$, an outer C^3-domain for $m = 1$, an outer C^4-domain for $m = 2$ and an outer $C^{2m-1,1}$-domain for $m \in \mathbb{N}$, $m \geq 3$. Furthermore let $(x_k)_{k \in \mathbb{N}}$ be a fundamental system in Σ or D. Then the system of mass point representations*

$$\left(\left. \frac{1}{|x_k - \cdot|} \right|_{\partial\Sigma} \right)_{k \in \mathbb{N}},$$

defined by Definition 2.6.1, is dense in $H^{m,2}(\partial\Sigma)$. Furthermore the system of outer harmonics

$$\left(\left. H^\gamma_{-n-1,k} \right|_{\partial\Sigma} \right)_{n=0,1,\ldots;k=1,\ldots,2n+1},$$

defined in Definition 2.6.2, is dense in $H^{m,2}(\partial\Sigma)$ if $0 < \gamma < \infty$ is such small that $B_\gamma^{\mathbb{R}^3}(0) \subset D$. Finally the system of inner harmonics

$$\left(\left. H^\gamma_{n,k} \right|_{\partial\Sigma} \right)_{n=0,1,\ldots;k=1,\ldots,2n+1},$$

defined in Definition 2.6.2, is dense in $H^{m,2}(\partial\Sigma)$ if $0 < \gamma < \infty$ is such large that $\overline{D} \subset B_\gamma^{\mathbb{R}^3}(0)$.

Proof. We want to use the fact that a subset S of a Hilbert space H is dense if and only if $\langle F, G \rangle_H = 0$ for all $G \in S$ yields $F = 0$ in H. Therefore let $\tilde{F} \in H^{m,2}(\partial\Sigma)$ be given. We have to show that

$$\left\langle \tilde{F}(\cdot), \left. \frac{1}{|x_k - \cdot|} \right|_{\partial\Sigma} \right\rangle_{H^{m,2}(\partial\Sigma)} = 0,$$

for all $k \in \mathbb{N}$, where $(x_k)_{k \in \mathbb{N}}$ is a fundamental system in D or Σ,

$$\left\langle \tilde{F}(\cdot), \left. H^\gamma_{-n-1,k} \right|_{\partial\Sigma} \right\rangle_{H^{m,2}(\partial\Sigma)} = 0,$$

or

$$\left\langle \tilde{F}(\cdot), \left. H^\gamma_{n,k} \right|_{\partial\Sigma} \right\rangle_{H^{m,2}(\partial\Sigma)} = 0,$$

for all $n = 0, 1, \ldots$ and $k = 1, \ldots, 2n+1$, yields $\tilde{F} = 0$ in $H^{m,2}(\partial\Sigma)$. We define the operator $I : H^{m,2} \to C^\infty(\mathbb{R}^3 \setminus \partial\Sigma)$ via

$$I[F](x) := \left\langle F(\,\cdot\,), \frac{1}{|x - \cdot|}\bigg|_{\partial\Sigma} \right\rangle_{H^{m,2}(\partial\Sigma)},$$

for all $F \in H^{m,2}(\partial\Sigma)$. We want to prove that $I[\tilde{F}](x) = 0$ for all $x \in D$. First consider the function system of mass point representations for a fundamental system $(x_k)_{k \in \mathbb{N}}$ in D. Obviously we have $I[\tilde{F}](x_k) = 0$ for all $k \in \mathbb{N}$, $I[\tilde{F}]$ is an analytic function on $\mathbb{R}^3 \setminus \partial\Sigma$ and $\Delta I[\tilde{F}](x) = 0$ for all $x \in \mathbb{R}^3 \setminus \partial\Sigma$. Remember that $\Delta \frac{1}{|x-y|} = 0$, for all $x \neq y$, and we are allowed to interchange the order of differentiation. Thus the fact that $(x_k)_{k \in \mathbb{N}}$ is a fundamental system in D yields $I[\tilde{F}](x) = 0$ for all $x \in D$. Now consider the function system of outer harmonics. Recall the series expansion from Lemma 2.6.3, given by

$$\frac{1}{|x-y|} = \sum_{n=0}^{\infty} \frac{4\pi\gamma}{2n+1} \sum_{k=1}^{2n+1} H_{n,j}^\gamma(x) H_{-n-1,j}^\gamma(y),$$

for $y \in \partial\Sigma$ and $x \in B_\gamma^{\mathbb{R}^3}(0) \subset D$. For fixed x the series converges uniformly on $\partial\Sigma$ in y, see [FM04, p.90]. Moreover, [Mar68, Satz 3.19] gives the convergence even in $C^n(\partial\Sigma)$ for arbitrary $n \in \mathbb{N}$. Thus

$$I[\tilde{F}](x) = \left\langle \tilde{F}(\,\cdot\,), \frac{1}{|x - \cdot|}\bigg|_{\partial\Sigma} \right\rangle_{H^{m,2}(\partial\Sigma)}$$

$$= \left\langle \tilde{F}(\,\cdot\,), \sum_{n=0}^{\infty} \frac{4\pi\gamma}{2n+1} \sum_{j=1}^{2n+1} H_{n,j}^\gamma(x) H_{-n-1,j}^\gamma(y) \right\rangle_{H^{m,2}(\partial\Sigma)}$$

$$= \sum_{n=0}^{\infty} \frac{4\pi\gamma}{2n+1} \sum_{j=1}^{2n+1} H_{n,j}^\gamma(x) \left\langle \tilde{F}(\,\cdot\,), H_{-n-1,j}^\gamma(y) \right\rangle_{H^{m,2}(\partial\Sigma)} = 0,$$

for all $x \in B_\gamma^{\mathbb{R}^3}(0) \subset D$, by the continuity of the scalar product. By analytic continuation we find $I[\tilde{F}](x) = 0$ on D, see again [FM04, p.91]. Analogously, we find $I[\tilde{F}](x) = 0$ on Σ for the inner harmonics or $\left(\frac{1}{|x_k-y|}\right)_{k \in \mathbb{N}}$ corresponding to a fundamental system in Σ. From now on the proof will not depend on the function system any more. We define $I^{\pm\tau} : H^{m,2}(\partial\Sigma) \to L^2(\partial\Sigma)$ for all $\tau \in (0, \tau_0]$ by

$$I^{\pm\tau}[F](x) := U[F](x - \tau\nu(x)),$$

for all $F \in H^{m,2}(\partial\Sigma)$. We have that $I^{\pm\tau}$ is linear and bounded for each $\tau \in (0, \tau_0]$

$$\|I^{\pm\tau}[F]\|_{L^2(\partial\Sigma)} = \|(F(y), \frac{1}{|x \pm \tau\nu(x) - y|})_{H^{m,2}(\partial\Sigma)}\|_{L^2(\partial\Sigma)}$$

4.7. APPLICATION OF THE LIMIT FORMULAE TO GEOMATHEMATICS

$$\leq \|F\|_{H^{m,2}(\partial\Sigma)} \left\| \left\| \frac{1}{|x \pm \tau\nu(x) - y|} \right\|_{H^{m,2}(\partial\Sigma)} \right\|_{L^2(\partial\Sigma)}$$

$$\leq \|F\|_{H^{m,2}(\partial\Sigma)} H^2(\partial\Sigma) \sup_{x \in \partial\Sigma} \left\| \frac{1}{|x \pm \tau\nu(x) - \cdot|} \right\|_{C^m(\partial\Sigma)}$$

$$= \|F\|_{H^{m,2}(\partial\Sigma)} H^2(\partial\Sigma) \underbrace{\sup_{z \in \partial\Sigma^{\pm\tau}} \left\| \frac{1}{|z - \cdot|} \right\|_{C^m(\partial\Sigma)}}_{<\infty}.$$

Now we can identify $I^{\pm\tau}$ with a linear continuous mapping from $H^{m,2}(\partial\Sigma)$ to $(H^{m,2}(\partial\Sigma))'$ with help of the embedding $L^2(\partial\Sigma) \to (H^{m,2}(\partial\Sigma))'$ in form of regular distributions. Furthermore we have for each $\tau \in (0, \tau_0]$ and all $F \in C^\infty(\partial\Sigma)$, that $I^{\pm\tau}[F] = U_1^{\pm\tau}[D^{2m}[F]]$ in sense of $(H^{m,2}(\partial\Sigma))'$ because

$$I^{\pm\tau}[F](G) = \langle \langle F(y), \frac{1}{|x \pm \tau\nu(x) - y|} \rangle_{H^{m,2}(\partial\Sigma)}, G(x) \rangle_{L^2(\partial\Sigma)}$$
$$= \langle \langle D^{2m}F(y), \frac{1}{|x \pm \tau\nu(x) - y|} \rangle_{L^2(\partial\Sigma)}, G(x) \rangle_{L^2(\partial\Sigma)}$$
$$= \langle U_1[D^{2m}F](x \pm \tau\nu(x)), G(x) \rangle_{L^2(\partial\Sigma)}$$
$$= \langle U_1^{\pm\tau}[D^{2m}F](x), G(x) \rangle_{L^2(\partial\Sigma)}$$
$$= \langle D^{2m}F(x), (U_1^{\pm\tau})^*[G](x) \rangle_{L^2(\partial\Sigma)}$$
$$= D^{2m}F((U_1^{\pm\tau})^*[G]) = U_1^{\pm\tau}[D^{2m}F](G),$$

see Lemma 4.7.6. As well $I^{\pm\tau}$ as $U_1^{\pm\tau} \circ D^{2m}$ are continuous operators from $H^{m,2}(\partial\Sigma)$ to $(H^{m,2}(\partial\Sigma))'$ and both coincide on the dense subset $C^\infty(\partial\Sigma) \subset H^{m,2}(\partial\Sigma)$. So we have that they are identical on all of $H^{m,2}(\partial\Sigma)$ and we find

$$U_1^{\pm\tau}[D^{2m}[\tilde{F}]] = I^{\pm\tau}[\tilde{F}] = 0,$$

in sense of $(H^{m,2}(\partial\Sigma))'$ for all $\tau \in (0, \tau_0]$. Using the limit formulae in $(H^{m,2}(\partial\Sigma))'$-norm, see Theorem 4.7.4, we get

$$\lim_{\tau \to 0^+} \|U_1^{\pm\tau}[D^{2m}[\tilde{F}]] - U_1^0[D^{2m}[\tilde{F}]]\|_{(H^{m,2}(\partial\Sigma))'} = \lim_{\tau \to 0^+} \|U_1^0[D^{2m}[\tilde{F}]]\|_{(H^{m,2}(\partial\Sigma))'} = 0.$$

This gives

$$D^{2m}[\tilde{F}]((U_1^0)^*[G]) = D^{2m}[\tilde{F}](U_1[G]) = 0,$$

for all $G \in H^{m,2}(\partial\Sigma)$, by the definition of $(U_1^0)^*$ and consequently

$$\langle \tilde{F}, U_1[G] \rangle_{H^{m,2}(\partial\Sigma)} = 0,$$

for all $G \in H^{m,2}(\partial\Sigma)$, by definition of the operator D^{2m}. Due to Lemma 4.7.7 we have that the range of $U_1 : H^{m,2}(\partial\Sigma) \to H^{m,2}(\partial\Sigma)$ contains a dense subset of $H^{m,2}(\partial\Sigma)$, namely $C^{m+2}(\partial\Sigma)$. Thus $\tilde{F} = 0$ in $H^{m,2}(\partial\Sigma)$. Consequently the proof is done. $\square$

4.8 Appendix: Oblique Limit Formulae

This section was added to my dissertation to complete the theory. In geomathematematical problems, oblique boundary problems frequently arise. The reason is that in general the normal vector of the earths surface and the gravity vector do not coincide. Therefore, oblique boundary conditions seem to be more natural then Neumann boundary conditions, because those are only applicable for rough approaches, e.g. assuming the earth as a sphere. For this reason we will translate the theory of limit formulae in this appendix to the case of an oblique approach to the surface. This might lead to density results for oblique derivatives of the function systems considered and thus be used to construct solutions to oblique problems. Aim of this appendix is to extend the results presented in the previous sections to the case when we approach from an oblique direction to the surface $\partial\Sigma$. We investigate the single layer potential U_1 as well as the oblique derivative of the single layer potential $\frac{\partial U_1}{\partial a}$. We start by introducing the oblique vector field a which forms the basis of the further considerations. For the rest of this final section we assume Σ to be at least an outer $C^{2,\alpha}$-domain, $0 < \alpha \leq 1$, and $a \in C^{1,\alpha}(\partial\Sigma; \mathbb{R}^3)$ such that the regularity condition

$$|(a(x), \nu(x))| > C_{19} > 0$$

holds on $\partial\Sigma$ with a constant $0 < C_{19} < \infty$ independent of x. For a constant $0 < \tau_1 < \infty$ we define the oblique parallel surfaces $\partial\Sigma_a^{\pm\tau}$ for each $\tau \in [0, \tau_1]$ by

$$\partial\Sigma_a^{\pm\tau} := \left\{ x \pm \tau a(x) \big| x \in \partial\Sigma \right\}.$$

We can choose τ_1 such small that $x \pm \tau a(x) \notin \partial\Sigma$ for all $x \in \partial\Sigma$ and $\tau \in (0, \tau_1]$. The oblique setting is illustrated in Figure 8.

4.8. APPENDIX: OBLIQUE LIMIT FORMULAE

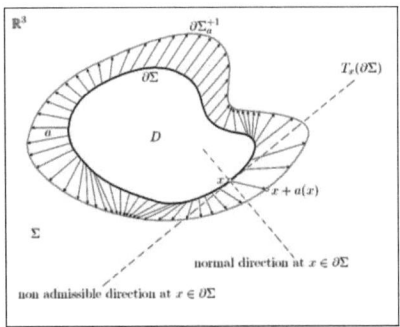

Figure 8: Oblique parallel surfaces

The evaluation on $\partial\Sigma_a^{\pm\tau}$ of the single layer potential as well as of the oblique derivative of the single layer potential is defined as follows.

Definition 4.8.1. Let Σ be an outer $C^{2,\alpha}$-domain, $0 < \alpha \leq 1$, $a \in C^{1,\alpha}(\partial\Sigma;\mathbb{R}^3)$ and $F \in C^{0,\alpha}(\partial\Sigma)$. Then we define

$$U_1(x \pm \tau a(x)) := \int_{\partial\Sigma} F(y) \frac{1}{|x \pm \tau a(x) - y|} \, dH^2(y),$$

$$\frac{\partial U_1}{\partial a}(x \pm \tau a(x)) := (a(x), (\nabla U_1)(x \pm \tau a(x))),$$

for all $\tau \in (0, \tau_1]$ and $x \in \partial\Sigma$. Moreover we set

$$U_1(x) := \int_{\partial\Sigma} F(y) \frac{1}{|x - y|} \, dH^2(y),$$

$$\frac{\partial U_1}{\partial a}(x) := \int_{\partial\Sigma} F(y) \left(a(x), \nabla_x \frac{1}{|x - y|} \right) dH^2(y),$$

for all $x \in \partial\Sigma$. The last integral exists as Cauchy principal value. Furthermore $\frac{1}{|x-y|}$ forms a weakly singular integral kernel and thus all the other integrals are well defined, too.

We have the following oblique limit formulae.

Lemma 4.8.2. Let Σ be an outer $C^{2,\alpha}$-domain, $0 < \alpha \leq 1$, $a \in C^{1,\alpha}(\partial\Sigma;\mathbb{R}^3)$ and $F \in C^{0,\alpha}(\partial\Sigma)$. The oblique limit formulae are given by

$$\lim_{\tau \to 0^+} U_1[F](x \pm \tau a(x)) = U_1[F](x) \quad \text{for all } x \in \partial\Sigma,$$

$$\lim_{\tau \to 0^+} \frac{\partial U_1}{\partial a}[F](x \pm \tau a(x)) = \frac{\partial U_1}{\partial a}[F](x) \mp 2\pi F(x)(a(x), \nu(x)) \quad \text{for all } x \in \partial\Sigma.$$

The convergence is even uniform in $x \in \partial\Sigma$ and holds also in $C^{0,\beta}(\partial\Sigma)$ for all $0 < \beta < \alpha$.

Proof. All results of this section presented until now are taken from [FM04, Para. 3.3.2]. □

Remark 4.8.3. Note that for $a = \nu$ we end up with the classical limit formulae, see Lemma 4.2.2. In [FM04], for an outer $C^{2,\alpha}$-domain, $0 < \alpha \leq 1$, $a \in C^{1,\alpha}(\partial\Sigma; \mathbb{R}^3)$ and $F \in C^{0,\alpha}(\partial\Sigma)$ also for the oblique double layer potential

$$U_2^a(x \pm \tau a(x)) := \int_{\partial\Sigma} F(y) \left(a(y), \nabla_y \frac{1}{|x \pm \tau a(x) - y|} \right) dH^2(y),$$

$x \in \partial\Sigma$ and $\tau \in [0, \tau_1]$, a limit formula is given. For $\tau = 0$ the integral again only exists as Cauchy principal value. We have

$$\lim_{\tau \to 0^+} U_2^a[F](x \pm \tau a(x)) = U_2^a[F](x) \pm 2\pi F(x)(a(x), \nu(x)) \quad \text{for all } x \in \partial\Sigma.$$

Again the convergence even holds in $C^{0,\beta}(\partial\Sigma)$ for all $0 \leq \beta < \alpha$. Because a result analogous to Lemma 4.3.1 for U_2^a is yet missing, we can not extend our theory for this potential and thus we restrict our investigation to U_1 and $\frac{\partial U_1}{\partial a}$.

Again we define a potential operator for each of the two limit formulae.

Lemma 4.8.4. Let Σ be an outer $C^{2,\alpha}$-domain, $0 < \alpha \leq 1$, $a \in C^{1,\alpha}(\partial\Sigma; \mathbb{R}^3)$ and $F \in C^{0,\alpha}(\partial\Sigma)$. We define the potential operators $T_1^{\pm\tau}[F]$ and $T_2^{\pm\tau}[F]$ by

$$T_1^{\pm\tau}[F](x) := U_1[F](x \pm \tau a(x)) - U_1[F](x),$$

$$T_2^{\pm\tau}[F](x) := \frac{\partial U_1}{\partial a}[F](x \pm \tau a(x)) - \frac{\partial U_1}{\partial a}[F](x) \pm 2\pi F(x)(a(x), \nu(x)),$$

for all $x \in \partial\Sigma$ and $\tau \in (0, \tau_1]$. We have

$$\lim_{\tau \to 0^+} T_i^{\pm\tau}[F](x) = 0, \qquad (4.45)$$

for all $x \in \partial\Sigma$ and $i = 1, 2$. Furthermore, we have $T_i^{\pm\tau}[F] \in C^{0,\beta}(\partial\Sigma)$ for $i = 1, 2$, all $0 \leq \beta < \alpha$ and all $\tau \in (0, \tau_1]$ with

$$\lim_{\tau \to 0^+} \|T_i^{\pm\tau}[F]\|_{C^{0,\beta}(\partial\Sigma)} = 0, \qquad (4.46)$$

for $i = 1, 2$.

Remark 4.8.5. Also in the oblique case we restrict our considerations to the limit formulae, while the jump relations drop as corollaries. Under the assumptions of Lemma 4.8.2 we have

$$\lim_{\tau \to 0^+} U_1[F](x + \tau a(x)) - U_1[F](x - \tau a(x)) = 0,$$

4.8. APPENDIX: OBLIQUE LIMIT FORMULAE

$$\lim_{\tau \to 0^+} U_1[F](x + \tau a(x)) + U_1[F](x - \tau a(x)) = 2U_1[F](x),$$

$$\lim_{\tau \to 0^+} \frac{\partial U_1}{\partial a}[F](x + \tau a(x)) - \frac{\partial U_1}{\partial a}[F](x - \tau a(x)) = -2\pi F(x)\langle a(x), \nu(x)\rangle,$$

$$\lim_{\tau \to 0^+} \frac{\partial U_1}{\partial a}[F](x + \tau a(x)) + \frac{\partial U_1}{\partial a}[F](x - \tau a(x)) = 2\frac{\partial U_1}{\partial a}[F](x),$$

for all $x \in \partial\Sigma$ and consequently also in $C^{0,\beta}(\partial\Sigma)$ for all $0 \leq \beta < \alpha$.

We start by translating Theorem 4.3.3 to the oblique case, i.e., the convergence in the function space $C^m(\partial\Sigma)$.

Theorem 4.8.6. *Let $m \in \mathbb{N}$ and $0 < \alpha \leq 1$ be given. For $m = 1$ let $a \in C^{1,\alpha}(\partial\Sigma; \mathbb{R}^3)$ and Σ be an outer $C^{2,\alpha}$-domain. Furthermore, let $F \in C^{0,\alpha}(\partial\Sigma)$ for $T_1^{\pm\tau}$ and $F \in C^{1,\alpha}(\partial\Sigma)$ for $T_2^{\pm\tau}$. For $m \geq 2$ let $a \in C^m(\partial\Sigma; \mathbb{R}^3)$, Σ be an outer $C^{m,\alpha}$-domain and $F \in C^{m-1,\alpha}(\partial\Sigma)$ for $T_1^{\pm\tau}$. Moreover, let Σ be an outer $C^{m+1,\alpha}$-domain and $F \in C^{m,\alpha}(\partial\Sigma)$ for $T_2^{\pm\tau}$. Then $T_i^{\pm\tau}[F] \in C^m(\partial\Sigma)$ for $i = 1, 2$, $\tau \in (0, \tau_1]$, and we have*

$$\lim_{\tau \to 0^+} \|T_i^{\pm\tau}[F]\|_{C^m(\partial\Sigma)} = 0. \tag{4.47}$$

Proof. Recall the proof of Theorem 4.3.3 and the result from Lemma 4.3.1. Because the proof of Theorem 4.8.6 differs only slightly, we only mention the necessary modifications. We use the notations from the proof of Theorem 4.3.3. Let the assumptions of the theorem be fulfilled. In the oblique case we define for $x \in \partial\Sigma$ and $\tau \in (0, \tau_1]$

$$V_{\pm\tau}^1[F](x) := \overline{U_1[F]}(x) - U_1[F](x \pm \tau a(x)),$$

$$V_{\pm\tau}^2[F](x) := \frac{\partial}{\partial a(x)}\overline{U_1[F]}(x) - \frac{\partial U_1}{\partial a(x)}[F](x \pm \tau a(x))$$

$$= (a(x), \nabla\overline{U_1[F]}(x) - \nabla U_1[F](x \pm \tau a(x))).$$

Here $\overline{U}$ denotes the continuation of the function U from Σ or D to an open set Σ_ϵ or D_ϵ containing $\overline{\Sigma}$ or $\overline{D}$, respectively. It holds $V_{\pm\tau}^i[F] \in C^m(\partial\Sigma)$. Because we can estimate

$$\|(a(\,\cdot\,), \nabla\overline{U_1[F]}(\,\cdot\,) - \nabla U_1[F](\,\cdot \pm \tau\nu(\,\cdot\,)))\|_{C^m(\partial\Sigma)}$$

$$\leq \|a\|_{C^m(\partial\Sigma)} \sum_{k=1}^{3} \|\partial_k\overline{U_1[F]}(\,\cdot\,) - \partial_k U_1[F](\,\cdot \pm \tau\nu(\,\cdot\,))\|_{C^m(\partial\Sigma)}.$$

it suffices to prove

$$\lim_{\tau \to 0^+} \|W_{\pm\tau}^{i,j,s}[F]\|_{C^0(B_1^{\mathbb{R}^2}(0))} = 0,$$

for $i = 1, 2$, $j = 1, \ldots, N$ and $1 \leq |s| \leq m$, with

$$W^{1,j,s}_{\pm\tau}[F](x) := \partial_1^{s_1} \partial_2^{s_2} \left(\overline{U_1[F]}(\Psi_j(x)) - U_1[F](\Psi_j(x) \pm \tau a(\Psi_j(x))) \right),$$

$$W^{2,j,s}_{\pm\tau}[F](x) := \partial_1^{s_1} \partial_2^{s_2} \left(\partial_k \overline{U_1[F]}(\Psi_j(x)) - \partial_k U_1[F](\Psi_j(x) \pm \tau a(\Psi_j(x))) \right),$$

for all $x \in B_1^{\mathbb{R}^2}(0)$, $\tau \in (0, \tau_1]$, $1 \leq s_1 + s_2 \leq m$, $k = 1, 2, 3$ and $j = 1, \ldots, N$. This can be done analogously to the normal setting using the uniform convergence of U_1 and its derivatives defined on Σ_ϵ or D_ϵ when approaching to the surface for τ tending to zero. The only difference is that we use regularity of the oblique vector field a instead of the normal vector field ν. This is also the reason why we can weaken the assumptions on the surface compared to Theorem 4.3.3. Thus the theorem is assumed to be proved. □

We proceed in this section by translating the convergence in $C^{m,\beta}(\partial\Sigma)$-norm from Section 5 to the oblique case. In order to prove it, we use Lemma 4.3.1 and Theorem 4.8.6.

Theorem 4.8.7. *Let $m \in \mathbb{N}$ and $0 < \beta < \alpha$ be given. For $m = 1$ let $a \in C^{1,\alpha}(\partial\Sigma; \mathbb{R}^3)$ and Σ be an outer $C^{2,\alpha}$-domain. Furthermore, let $F \in C^{0,\alpha}(\partial\Sigma)$ for $T_1^{\pm\tau}$ and $F \in C^{1,\alpha}(\partial\Sigma)$ for $T_2^{\pm\tau}$. For $m \geq 2$ let $a \in C^{m,\beta}(\partial\Sigma; \mathbb{R}^3)$, Σ be an outer $C^{m,\alpha}$-domain and $F \in C^{m-1,\alpha}(\partial\Sigma)$ for $T_1^{\pm\tau}$. Moreover, let Σ be an outer $C^{m+1,\alpha}$-domain and $F \in C^{m,\alpha}(\partial\Sigma)$ for $T_2^{\pm\tau}$. Then we have $T_i^{\pm\tau}[F] \in C^{m,\beta}(\partial\Sigma)$ for all $\tau \in (0, \tau_1]$, $i = 1, 2$ and*

$$\lim_{\tau \to 0^+} \|T_i^{\pm\tau}[F]\|_{C^{m,\beta}(\partial\Sigma)} = 0. \tag{4.48}$$

Proof. Again the proof runs in accordance to the normal setting, see Theorem 4.4.1. Also in the oblique case we find due to the assumptions $U_1[F] \in C^{m,\beta}(D)$, $U_1[F] \in C^{m,\beta}(\Sigma)$, for $i = 1$, and $\nabla U_1[F] \in C^{m,\beta}(D)$, $\nabla U_1[F] \in C^{m,\beta}(\Sigma)$, for $i = 2$. Furthermore, Lemma 2.2.10 gives continuations $U_1[F] \in C^{m,\beta}(D_\epsilon)$, $U_1[F] \in C^{m,\beta}(\Sigma_\epsilon)$ and $\nabla U_1[F] \in C^{m,\beta}(D_\epsilon)$, $\nabla U_1[F] \in C^{m,\beta}(\Sigma_\epsilon)$, for an $0 < \epsilon < \infty$, where we kept the notation for D_ϵ and Σ_ϵ from the previous proof. We have for $x, y \in \partial\Sigma$

$$0 < \frac{|x \pm \tau a(x) - (y \pm \tau a(y))|}{|x - y|^\alpha} \leq \frac{|x - y|}{|x - y|^\alpha} + \tau_1 \frac{|a(x) - a(y)|}{|x - y|^\alpha} \leq 2\max_{\partial\Sigma} |x| + \text{höl}_\alpha(a) < \infty$$

because $a \in C^{0,\alpha}(\partial\Sigma)$, see Lemma 2.2.9, and by the choice of τ_1. We used the alternative definition of the Hölder constant introduced in the proof of Theorem 4.4.1. This yields

$$|x - y|^\alpha > C_{20}|x \pm \tau a(x) - (y \pm \tau a(y))|$$

4.8. APPENDIX: OBLIQUE LIMIT FORMULAE

for a constant $0 < C_{20} < \infty$. Thus we find $U_1[F](\,\cdot\, \pm \tau a(\,\cdot\,)) \in C^{m,\beta}(\partial\Sigma)$ for all $\tau \in [0, \tau_1]$, because we can estimate the Hölder constant by those of $U_1[F] \in C^{m,\beta}(D_\epsilon)$ or $U_1[F] \in C^{m,\beta}(\Sigma_\epsilon)$, respectively. We obtain in a similar way that $\partial_k U_1[F](\,\cdot\, \pm \tau a(\,\cdot\,)) \in C^{m,\beta}(\partial\Sigma)$, $k = 1, 2, 3$, $\tau \in [0, \tau_1]$. Going on in the proof, we estimate

$$\|(a(\,\cdot\,), \nabla\overline{U_1[F]}(\,\cdot\,) - \nabla U_1[F](\,\cdot\, \pm \tau a(\,\cdot\,)))\|_{C^{m,\beta}(\partial\Sigma)}$$

$$\leq c_4^1 c_4^2 \|a\|_{C^{m,\beta}(\partial\Sigma)} \sum_{j=1}^{3} \|\partial_j \overline{U_1[F]}(\,\cdot\,) - \partial_j U_1[F](\,\cdot\, \pm \tau a(\,\cdot\,))\|_{C^{m,\beta}(\partial\Sigma)}$$

with the constants from Lemma 2.2.9 and consequently, using the results of Theorem 4.8.6 it is left to proof

$$\lim_{\tau \to 0} \text{höl}_\beta(W_{\pm\tau}^{i,j,s}[F]) = 0,$$

for all $|s| = m$, $i = 1, 2$ and $j = 1, \ldots, N$, taking the definitions from the previous proof. This can be done in the same manner as for the normal approach, see Theorem 4.4.1. The only difference is that we use the oblique instead of the normal direction to approach the surface. Since this does not change the uniform convergence of $\tilde{U}$, see the proof of Theorem 4.4.1, for τ tending to zero, we refer to the proof of Theorem 4.4.1 and the theorem is assumed to be proved. □

We close the book with a final remark.

Remark 4.8.8. At the moment we are not able to translate the convergence results in $L^2(\partial\Sigma)$ and $H^{m,2}(\partial\Sigma)$ for the oblique setting. Non the less we want to mention that there is only one gap to close. Assume we are able to show that for an outer $C^{k,\delta}$-domain Σ, $k \in \mathbb{N}_0$, $0 \leq \delta \leq 1$, we have

$$\|U_1(\,\cdot\, \pm \tau a(\,\cdot\,))\|_{L^2(\partial\Sigma)} \leq C_{21} \|F\|_{H^{l,2}(\partial\Sigma)},$$

$$\|\frac{\partial U_1}{\partial a}(\,\cdot\, \pm \tau a(\,\cdot\,))\|_{L^2(\partial\Sigma)} \leq C_{21} \|F\|_{H^{l,2}(\partial\Sigma)},$$

for all $F \in C^\infty(\partial\Sigma)$, all $\tau \in [0, \tau_1]$, some $l \in \mathbb{N}_0$ and a constant $0 < C_{21} < \infty$ independent of τ. This is an result corresponding to Lemma 4.5.2 in the normal case. Then the proof of Theorem 4.6.1 translates to the oblique setting and we get the convergence of the oblique limit formulae in the spaces $H^{m_1,2}(\partial\Sigma)$ for $F \in H^{m_1,2}(\partial\Sigma)$, $m_1, m_2 \in \mathbb{N}_0$, $m_2 \geq m_1$. A basis to tackle this problem might be found in [FM04]. This could be useful to implement applications for solutions to oblique boundary problems. For more about this problem see e.g. [GR10a].

Bibliography

[Ada75] R. A. Adams. *Sobolev spaces*. Pure and Applied Mathematics, 65. A Series of Monographs and Textbooks. New York-San Francisco-London: Academic Press, Inc., 1975.

[Alt02] H. W. Alt. *Lineare Funktionalanalysis. Eine anwendungsorientierte Einführung. 4., überarb. und erweiterte Aufl.* Berlin: Springer, 2002.

[Alt04] H. W. Alt. *Lecture Notes on Analysis 3.* University of Bonn, 2004.

[Bau02] H. Bauer. *Wahrscheinlichkeitstheorie. 5., durchges. u. verbess. Aufl.* Berlin: de Gruyter, 2002.

[Bau04] F. Bauer. *An Alternative Approach to the Oblique Derivative Problem in Potential Theory.* Berichte aus der Mathematik. Aachen: Shaker Verlag; Kaiserslautern: Univ. Kaiserslautern, Fachbereich Mathematik (Diss. 2004), 2004.

[BK95] Yu. M. Berezanskij and Yu. G. Kondratiev. *Spectral methods in infinite-dimensional analysis. Vol. 1, 2.* Mathematical Physics and Applied Mathematics. 12. Dordrecht: Kluwer Academic Publishers, 1995.

[CK83] D. Colton and R. Kress. *Integral equation methods in scattering theory.* Pure and Applied Mathematics. A Wiley-Interscience Publication. New York etc.: John Wiley & Sons, 1983.

[DL88] R. Dautray and J.-L. Lions. *Mathematical analysis and numerical methods for science and technology. (In six volumes). Volume 2: Functional and variational methods.* With the collaboration of Michel Artola, Marc Authier, Philippe Bénilan, Michel Cessenat, Jean-Michel Combes, Hélène Lanchon, Bertrand Mercier, Claue Wild, Claude Zuily. Transl. from the French by Ian N. Sneddon. Berlin etc.: Springer-Verlag, 1988.

[Dob06] M. Dobrowolski. *Angewandte Funktionalanalysis. Funktionalanalysis, Sobolev-Räume und elliptische Differentialgleichungen.* Berlin: Springer, 2006.

[Fic48] G. Fichera. Teoremi di completezza sulla frontiera di un dominio per taluni sistemi di funzioni. *Ann. Mat. Pura Appl., IV. Ser.*, 27:1–28, 1948.

[Fis03] G. Fischer. *Lineare Algebra. Eine Einführung für Studienanfänger. 14., durchgesehene Aufl.* Wiesbaden: Vieweg, 2003.

[FK80] W. Freeden and H. Kersten. The geodetic boundary-value problem using the known surface of the earth. 1980.

[FM02] W. Freeden and T. Maier. On multiscale denoising of spherical functions: basic theory and numerical aspects. *ETNA, Electron. Trans. Numer. Anal.*, 14:56–78, 2002.

[FM03] W. Freeden and C. Mayer. Wavelets generated by layer potentials. *Appl. Comput. Harmon. Anal.*, 14(3):195–237, 2003.

[FM04] W. Freeden and V. Michel. *Multiscale potential theory. With applications to geoscience.* Applied and Numerical Harmonic Analysis. Boston, MA: Birkhäuser, 2004.

[For01] O. Forster. *Analysis 1. Differential- und Integralrechnung einer Veränderlichen. 6., verb. Aufl.* Braunschweig: Vieweg, 2001.

[For05] O. Forster. *Analysis 2. Differentialrechnung im $\mathbb{R}^n$, gewöhnliche Differentialgleichungen. 6th revised and enlarged ed.* Vieweg Studium: Grundkurs Mathematik. Wiesbaden: Vieweg, 2005.

[For07] O. Forster. *Analysis 3: Integralrechnung im $\mathbb{R}^n$ mit Anwendungen. 4th ed.* Vieweg Studium: Aufbaukurs Mathematik. Wiesbaden: Vieweg, 2007.

[Fre80] W. Freeden. On the Approximation of External Gravitational Potential with Closed Systems of (Trial) Functions. *Bull. Géod.*, 54:1–20, 1980.

[Geh70] H. W. Gehm. *Über vollständige und abgeschlossene Funktionensysteme auf regulären Flächen mit Anwendungen auf die Randwertprobleme der Helmholtzschen Schwingungsgleichung.* PhD thesis, Aachen: RWTH Aachen, Fachbereich Mathematik (Diss. 1970), 1970.

BIBLIOGRAPHY

[GR06] M. Grothaus and T. Raskop. On the oblique boundary problem with a stochastic inhomogeneity. *Stochastics*, 78(4):233–257, 2006.

[GR09] M. Grothaus and T. Raskop. The Outer Oblique Boundary Problem of Potential Theory. *Numer. Funct. Anal. Optim.*, 30(7-8):711–750, 2009.

[GR10a] M. Grothaus and T. Raskop. Limit Formulae and Jump Relations of Potential Theory in Sobolev Spaces. *accepted for publication in: GEM - International Journal on Geomathematics*, 1(1), 2010.

[GR10b] M. Grothaus and T. Raskop. Oblique Stochastic Boundary-Value-Problem. *accepted for publication in: Handbook of Geomathematics, Berlin: Springer*, 2010.

[GT01] D. Gilbarg and N. S. Trudinger. *Elliptic partial differential equations of second order. Reprint of the 1998 ed.* Classics in Mathematics. Berlin: Springer, 2001.

[Gün57] N.M. Günter. *Die Potentialtheorie und ihre Anwendung auf Grundaufgaben der mathematischen Physik.* Leipzig: B. G. Teubner Verlagsgesellschaft, 1957.

[Gut08] M. Gutting. *Fast multipole methods for oblique derivative problems.* Berichte aus der Mathematik. Aachen: Shaker Verlag; Kaiserslautern: Univ. Kaiserslautern, Fachbereich Mathematik (Diss. 2007), 2008.

[Haz97] M. Hazewinkel. *Encyclopaedia of mathematics. 10-volume English edition.* Dordrecht: Kluwer Academic Publishers, 1997.

[Heu01] H. Heuser. *Lehrbuch der Analysis. Teil 1. 14. Aufl.* Mathematische Leitfäden. Stuttgart: B.G. Teubner, 2001.

[Heu02] H. Heuser. *Lehrbuch der Analysis. Teil 2. 12. Aufl.* Stuttgart: B.G. Teubner, 2002.

[Jän01] K. Jänich. *Topology. (Topologie.) 7. Aufl.* Springer-Lehrbuch. Berlin: Springer, 2001.

[Kel67] O. D. Kellogg. *Foundations of potential theory.* Berlin-Heidelberg-New York: Springer-Verlag, 1967.

[Ker80] H. Kersten. Grenz- und Sprungrelationen für Potentiale mit quadratsummierbarer Flächenbelegung. *Result. Math.*, 3:17–24, 1980.

[Mar68] E. Martensen. *Potentialtheorie.* Leitfäden der angewandten Mathematik und Mechanik. 12. Stuttgart: B.G. Teubner, 1968.

[Mic72] S.G. Michlin. *Lehrgang der mathematischen Physik. In deutscher Sprache herausgegeben von F. Kuhnert und S. Prößdorf. Übersetzung aus dem Russischen von Siegfried Prößdorf und Bernd Silbermann.* Mathematische Lehrbücher und Monographien. I. Abt. Band XV. Berlin: Akademie-Verlag, 1972.

[Mir70] C. Miranda. *Partial differential equations of elliptic type.* Berlin-Heidelberg-New York: Springer-Verlag, 1970.

[Mül51] C. Müller. Die Potentiale einfacher und mehrfacher Flächenbelegungen. *Math. Ann.*, 123:235–262, 1951.

[Mül69] C. Müller. *Foundations of the mathematical theory of electromagnetic waves.* Die Grundlagen der mathematischen Wissenschaften. 155. Berlin-Heidelberg-New York: Springer-Verlag, 1969.

[Oma09] M. Omari. Reduction methods for layer potentials and their applications to limit formula in potential theory. Master's thesis, University of Kaiserslautern, 2009.

[Poi99] H. Poincaré. *Théorie du potentiel newtonien. Leçons professées à la Sorbonne pendant le premier semestre 1894-95. Rédigées par Ed. Le Roy, G. Vincent.* Paris: G. Carré et C. Naud. 366 S. gr. 8°. [Nature 60, 410.] , 1899.

[Ras05] T. Raskop. On the oblique boundary problem with a stochastic inhomogeneity. Master's thesis, University of Kaiserslautern, 2005.

[Ras09] T. Raskop. *The Analysis of Oblique Boundary Problems and Limit Formulae Motivated by Problems from Geomathematics.* PhD thesis, Kaiserslautern: TU Kaiserslautern, Fachbereich Mathematik (Diss. 2009), 2009.

[RS72] M. Reed and B. Simon. *Methods of modern mathematical physics. I: Functional analysis.* New York-London: Academic Press, Inc., 1972.

[RS75] M. Reed and B. Simon. *Methods of modern mathematical physics. II: Fourier analysis, self- adjointness.* New York - San Francisco - London: Academic Press, a subsidiary of Harcourt Brace Jovanovich, Publishers, 1975.

[RS01] Y. Rozanov and F. Sansò. The analysis of the Neumann and the oblique derivative problem: the theory of regularization and its stochastic version. *J. Geod.*, 75(7-8):391–398, 2001.

BIBLIOGRAPHY

[RS02a] Y. Rozanov and F. Sansò. On the stochastic versions of Neumann and oblique derivative problems. *Stochastics Stochastics Rep.*, 74(1-2):371–391, 2002.

[RS02b] Y. Rozanov and F. Sansò. The Analysis of the Neumann and Oblique Derivative Problem: Weak Theory. In Krumm P. Grafarend, E. and Schwarze, editors, *Geodesy: Challenge of the 3rd Millenium*. New York: Springer, 2002.

[Sch14] E. Schmidt. Bemerkungen zur Potentialtheorie. 1914.

[Sch31a] J. Schauder. Bemerkung zu meiner Arbeit "Potentialtheoretische Untersuchungen I (Anhang)". 1931.

[Sch31b] J. Schauder. Potentialtheoretische Untersuchungen. *I. M. Z.*, 33:602–640, 1931.

[SR91] X. Saint Raymond. *Elementary introduction to the theory of pseudodifferential operators*. Studies in Advanced Mathematics. Boca Raton, FL: CRC Press, 1991.

[Tes78] Horst Teschke. *Über die Darstellung harmonischer und metaharmonischer Funktionen in Gebieten mit nichtglatten Rändern*. PhD thesis, Aachen: RWTH Aachen, Fachbereich Mathematik (Diss. 1978), 1978.

[Wal71] W. Walter. *Einführung in die Potentialtheorie*. Mannheim-Vienna-Zürich: Bibliographisches Institut, 1971.

[Wer02] D. Werner. *Funktionalanalysis. 4., überarbeitete Aufl.* Springer Lehrbuch. Berlin: Springer, 2002.

Die VDM Verlagsservicegesellschaft sucht für wissenschaftliche Verlage abgeschlossene und herausragende

Dissertationen, Habilitationen, Diplomarbeiten, Master Theses, Magisterarbeiten usw.

für die kostenlose Publikation als Fachbuch.

Sie verfügen über eine Arbeit, die hohen inhaltlichen und formalen Ansprüchen genügt, und haben Interesse an einer honorarvergüteten Publikation?

Dann senden Sie bitte erste Informationen über sich und Ihre Arbeit per Email an *info@vdm-vsg.de*.

Sie erhalten kurzfristig unser Feedback!

VDM Verlagsservicegesellschaft mbH
Dudweiler Landstr. 99   Telefon +49 681 3720 174
D - 66123 Saarbrücken   Fax   +49 681 3720 1749
www.vdm-vsg.de

Die VDM Verlagsservicegesellschaft mbH vertritt

Printed by Books on Demand GmbH, Norderstedt / Germany